Cognitive Neuroscience: A Very Short Introduction

VERY SHORT INTRODUCTIONS are for anyone wanting a stimulating and accessible way into a new subject. They are written by experts, and have been translated into more than 45 different languages.

The series began in 1995, and now covers a wide variety of topics in every discipline. The VSI library now contains over 500 volumes—a Very Short Introduction to everything from Psychology and Philosophy of Science to American History and Relativity—and continues to grow in every subject area.

Titles in the series include the following:

ACCOUNTING Christopher Nobes
ADOLESCENCE Peter K. Smith
ADVERTISING Winston Fletcher
AFRICAN AMERICAN RELIGION Eddie S. Glaude Jr
AFRICAN HISTORY John Parker and Richard Rathbone
AFRICAN RELIGIONS Jacob K. Olupona
AGEING Nancy A. Pachana
AGNOSTICISM Robin Le Poidevin
AGRICULTURE Paul Brassley and Richard Soffe
ALEXANDER THE GREAT Hugh Bowden
ALGEBRA Peter M. Higgins
AMERICAN HISTORY Paul S. Boyer
AMERICAN IMMIGRATION David A. Gerber
AMERICAN LEGAL HISTORY G. Edward White
AMERICAN POLITICAL HISTORY Donald Critchlow
AMERICAN POLITICAL PARTIES AND ELECTIONS L. Sandy Maisel
AMERICAN POLITICS Richard M. Valelly
THE AMERICAN PRESIDENCY Charles O. Jones
THE AMERICAN REVOLUTION Robert J. Allison
AMERICAN SLAVERY Heather Andrea Williams
THE AMERICAN WEST Stephen Aron
AMERICAN WOMEN'S HISTORY Susan Ware
ANAESTHESIA Aidan O'Donnell
ANARCHISM Colin Ward
ANCIENT ASSYRIA Karen Radner
ANCIENT EGYPT Ian Shaw
ANCIENT EGYPTIAN ART AND ARCHITECTURE Christina Riggs
ANCIENT GREECE Paul Cartledge
THE ANCIENT NEAR EAST Amanda H. Podany
ANCIENT PHILOSOPHY Julia Annas
ANCIENT WARFARE Harry Sidebottom
ANGELS David Albert Jones
ANGLICANISM Mark Chapman
THE ANGLO-SAXON AGE John Blair
ANIMAL BEHAVIOUR Tristram D. Wyatt
THE ANIMAL KINGDOM Peter Holland
ANIMAL RIGHTS David DeGrazia
THE ANTARCTIC Klaus Dodds
ANTISEMITISM Steven Beller
ANXIETY Daniel Freeman and Jason Freeman
THE APOCRYPHAL GOSPELS Paul Foster
ARCHAEOLOGY Paul Bahn
ARCHITECTURE Andrew Ballantyne
ARISTOCRACY William Doyle
ARISTOTLE Jonathan Barnes
ART HISTORY Dana Arnold
ART THEORY Cynthia Freeland

ASIAN AMERICAN HISTORY Madeline Y. Hsu
ASTROBIOLOGY David C. Catling
ASTROPHYSICS James Binney
ATHEISM Julian Baggini
THE ATMOSPHERE Paul I. Palmer
AUGUSTINE Henry Chadwick
AUSTRALIA Kenneth Morgan
AUTISM Uta Frith
THE AVANT GARDE David Cottington
THE AZTECS Davíd Carrasco
BABYLONIA Trevor Bryce
BACTERIA Sebastian G. B. Amyes
BANKING John Goddard and John O. S. Wilson
BARTHES Jonathan Culler
THE BEATS David Sterritt
BEAUTY Roger Scruton
BEHAVIOURAL ECONOMICS Michelle Baddeley
BESTSELLERS John Sutherland
THE BIBLE John Riches
BIBLICAL ARCHAEOLOGY Eric H. Cline
BIOGRAPHY Hermione Lee
BLACK HOLES Katherine Blundell
BLOOD Chris Cooper
THE BLUES Elijah Wald
THE BODY Chris Shilling
THE BOOK OF MORMON Terryl Givens
BORDERS Alexander C. Diener and Joshua Hagen
THE BRAIN Michael O'Shea
THE BRICS Andrew F. Cooper
THE BRITISH CONSTITUTION Martin Loughlin
THE BRITISH EMPIRE Ashley Jackson
BRITISH POLITICS Anthony Wright
BUDDHA Michael Carrithers
BUDDHISM Damien Keown
BUDDHIST ETHICS Damien Keown
BYZANTIUM Peter Sarris
CALVINISM Jon Balserak
CANCER Nicholas James
CAPITALISM James Fulcher
CATHOLICISM Gerald O'Collins
CAUSATION Stephen Mumford and Rani Lill Anjum
THE CELL Terence Allen and Graham Cowling
THE CELTS Barry Cunliffe
CHAOS Leonard Smith
CHEMISTRY Peter Atkins
CHILD PSYCHOLOGY Usha Goswami
CHILDREN'S LITERATURE Kimberley Reynolds
CHINESE LITERATURE Sabina Knight
CHOICE THEORY Michael Allingham
CHRISTIAN ART Beth Williamson
CHRISTIAN ETHICS D. Stephen Long
CHRISTIANITY Linda Woodhead
CIRCADIAN RHYTHMS Russell Foster and Leon Kreitzman
CITIZENSHIP Richard Bellamy
CIVIL ENGINEERING David Muir Wood
CLASSICAL LITERATURE William Allan
CLASSICAL MYTHOLOGY Helen Morales
CLASSICS Mary Beard and John Henderson
CLAUSEWITZ Michael Howard
CLIMATE Mark Maslin
CLIMATE CHANGE Mark Maslin
CLINICAL PSYCHOLOGY Susan Llewelyn and Katie Aafjes-van Doorn
COGNITIVE NEUROSCIENCE Richard Passingham
THE COLD WAR Robert McMahon
COLONIAL AMERICA Alan Taylor
COLONIAL LATIN AMERICAN LITERATURE Rolena Adorno
COMBINATORICS Robin Wilson
COMEDY Matthew Bevis
COMMUNISM Leslie Holmes
COMPLEXITY John H. Holland
THE COMPUTER Darrel Ince
COMPUTER SCIENCE Subrata Dasgupta
CONFUCIANISM Daniel K. Gardner
THE CONQUISTADORS Matthew Restall and Felipe Fernández-Armesto
CONSCIENCE Paul Strohm
CONSCIOUSNESS Susan Blackmore
CONTEMPORARY ART Julian Stallabrass
CONTEMPORARY FICTION Robert Eaglestone
CONTINENTAL PHILOSOPHY Simon Critchley
COPERNICUS Owen Gingerich
CORAL REEFS Charles Sheppard

CORPORATE SOCIAL RESPONSIBILITY Jeremy Moon
CORRUPTION Leslie Holmes
COSMOLOGY Peter Coles
CRIME FICTION Richard Bradford
CRIMINAL JUSTICE Julian V. Roberts
CRITICAL THEORY Stephen Eric Bronner
THE CRUSADES Christopher Tyerman
CRYPTOGRAPHY Fred Piper and Sean Murphy
CRYSTALLOGRAPHY A. M. Glazer
THE CULTURAL REVOLUTION Richard Curt Kraus
DADA AND SURREALISM David Hopkins
DANTE Peter Hainsworth and David Robey
DARWIN Jonathan Howard
THE DEAD SEA SCROLLS Timothy Lim
DECOLONIZATION Dane Kennedy
DEMOCRACY Bernard Crick
DEPRESSION Jan Scott and Mary Jane Tacchi
DERRIDA Simon Glendinning
DESCARTES Tom Sorell
DESERTS Nick Middleton
DESIGN John Heskett
DEVELOPMENTAL BIOLOGY Lewis Wolpert
THE DEVIL Darren Oldridge
DIASPORA Kevin Kenny
DICTIONARIES Lynda Mugglestone
DINOSAURS David Norman
DIPLOMACY Joseph M. Siracusa
DOCUMENTARY FILM Patricia Aufderheide
DREAMING J. Allan Hobson
DRUGS Les Iversen
DRUIDS Barry Cunliffe
EARLY MUSIC Thomas Forrest Kelly
THE EARTH Martin Redfern
EARTH SYSTEM SCIENCE Tim Lenton
ECONOMICS Partha Dasgupta
EDUCATION Gary Thomas
EGYPTIAN MYTH Geraldine Pinch
EIGHTEENTH-CENTURY BRITAIN Paul Langford
THE ELEMENTS Philip Ball
EMOTION Dylan Evans
EMPIRE Stephen Howe
ENGELS Terrell Carver
ENGINEERING David Blockley
ENGLISH LITERATURE Jonathan Bate
THE ENLIGHTENMENT John Robertson
ENTREPRENEURSHIP Paul Westhead and Mike Wright
ENVIRONMENTAL ECONOMICS Stephen Smith
ENVIRONMENTAL POLITICS Andrew Dobson
EPICUREANISM Catherine Wilson
EPIDEMIOLOGY Rodolfo Saracci
ETHICS Simon Blackburn
ETHNOMUSICOLOGY Timothy Rice
THE ETRUSCANS Christopher Smith
EUGENICS Philippa Levine
THE EUROPEAN UNION John Pinder and Simon Usherwood
EVOLUTION Brian and Deborah Charlesworth
EXISTENTIALISM Thomas Flynn
EXPLORATION Stewart A. Weaver
THE EYE Michael Land
FAMILY LAW Jonathan Herring
FASCISM Kevin Passmore
FASHION Rebecca Arnold
FEMINISM Margaret Walters
FILM Michael Wood
FILM MUSIC Kathryn Kalinak
THE FIRST WORLD WAR Michael Howard
FOLK MUSIC Mark Slobin
FOOD John Krebs
FORENSIC PSYCHOLOGY David Canter
FORENSIC SCIENCE Jim Fraser
FORESTS Jaboury Ghazoul
FOSSILS Keith Thomson
FOUCAULT Gary Gutting
THE FOUNDING FATHERS R. B. Bernstein
FRACTALS Kenneth Falconer
FREE SPEECH Nigel Warburton
FREE WILL Thomas Pink
FRENCH LITERATURE John D. Lyons
THE FRENCH REVOLUTION William Doyle
FREUD Anthony Storr
FUNDAMENTALISM Malise Ruthven

FUNGI Nicholas P. Money
THE FUTURE Jennifer M. Gidley
GALAXIES John Gribbin
GALILEO Stillman Drake
GAME THEORY Ken Binmore
GANDHI Bhikhu Parekh
GENES Jonathan Slack
GENIUS Andrew Robinson
GEOGRAPHY John Matthews and David Herbert
GEOPOLITICS Klaus Dodds
GERMAN LITERATURE Nicholas Boyle
GERMAN PHILOSOPHY Andrew Bowie
GLOBAL CATASTROPHES Bill McGuire
GLOBAL ECONOMIC HISTORY Robert C. Allen
GLOBALIZATION Manfred Steger
GOD John Bowker
GOETHE Ritchie Robertson
THE GOTHIC Nick Groom
GOVERNANCE Mark Bevir
GRAVITY Timothy Clifton
THE GREAT DEPRESSION AND THE NEW DEAL Eric Rauchway
HABERMAS James Gordon Finlayson
THE HABSBURG EMPIRE Martyn Rady
HAPPINESS Daniel M. Haybron
THE HARLEM RENAISSANCE Cheryl A. Wall
THE HEBREW BIBLE AS LITERATURE Tod Linafelt
HEGEL Peter Singer
HEIDEGGER Michael Inwood
HERMENEUTICS Jens Zimmermann
HERODOTUS Jennifer T. Roberts
HIEROGLYPHS Penelope Wilson
HINDUISM Kim Knott
HISTORY John H. Arnold
THE HISTORY OF ASTRONOMY Michael Hoskin
THE HISTORY OF CHEMISTRY William H. Brock
THE HISTORY OF LIFE Michael Benton
THE HISTORY OF MATHEMATICS Jacqueline Stedall
THE HISTORY OF MEDICINE William Bynum
THE HISTORY OF TIME Leofranc Holford-Strevens
HIV AND AIDS Alan Whiteside
HOBBES Richard Tuck
HOLLYWOOD Peter Decherney
HOME Michael Allen Fox
HORMONES Martin Luck
HUMAN ANATOMY Leslie Klenerman
HUMAN EVOLUTION Bernard Wood
HUMAN RIGHTS Andrew Clapham
HUMANISM Stephen Law
HUME A. J. Ayer
HUMOUR Noël Carroll
THE ICE AGE Jamie Woodward
IDEOLOGY Michael Freeden
INDIAN CINEMA Ashish Rajadhyaksha
INDIAN PHILOSOPHY Sue Hamilton
THE INDUSTRIAL REVOLUTION Robert C. Allen
INFECTIOUS DISEASE Marta L. Wayne and Benjamin M. Bolker
INFINITY Ian Stewart
INFORMATION Luciano Floridi
INNOVATION Mark Dodgson and David Gann
INTELLIGENCE Ian J. Deary
INTELLECTUAL PROPERTY Siva Vaidhyanathan
INTERNATIONAL LAW Vaughan Lowe
INTERNATIONAL MIGRATION Khalid Koser
INTERNATIONAL RELATIONS Paul Wilkinson
INTERNATIONAL SECURITY Christopher S. Browning
IRAN Ali M. Ansari
ISLAM Malise Ruthven
ISLAMIC HISTORY Adam Silverstein
ISOTOPES Rob Ellam
ITALIAN LITERATURE Peter Hainsworth and David Robey
JESUS Richard Bauckham
JOURNALISM Ian Hargreaves
JUDAISM Norman Solomon
JUNG Anthony Stevens
KABBALAH Joseph Dan
KAFKA Ritchie Robertson
KANT Roger Scruton
KEYNES Robert Skidelsky
KIERKEGAARD Patrick Gardiner
KNOWLEDGE Jennifer Nagel

THE KORAN Michael Cook
LANDSCAPE ARCHITECTURE
Ian H. Thompson
LANDSCAPES AND GEOMORPHOLOGY
Andrew Goudie and Heather Viles
LANGUAGES Stephen R. Anderson
LATE ANTIQUITY Gillian Clark
LAW Raymond Wacks
THE LAWS OF THERMODYNAMICS
Peter Atkins
LEADERSHIP Keith Grint
LEARNING Mark Haselgrove
LEIBNIZ Maria Rosa Antognazza
LIBERALISM Michael Freeden
LIGHT Ian Walmsley
LINCOLN Allen C. Guelzo
LINGUISTICS Peter Matthews
LITERARY THEORY Jonathan Culler
LOCKE John Dunn
LOGIC Graham Priest
LOVE Ronald de Sousa
MACHIAVELLI Quentin Skinner
MADNESS Andrew Scull
MAGIC Owen Davies
MAGNA CARTA Nicholas Vincent
MAGNETISM Stephen Blundell
MALTHUS Donald Winch
MANAGEMENT John Hendry
MAO Delia Davin
MARINE BIOLOGY Philip V. Mladenov
THE MARQUIS DE SADE John Phillips
MARTIN LUTHER Scott H. Hendrix
MARTYRDOM Jolyon Mitchell
MARX Peter Singer
MATERIALS Christopher Hall
MATHEMATICS Timothy Gowers
THE MEANING OF LIFE
Terry Eagleton
MEASUREMENT David Hand
MEDICAL ETHICS Tony Hope
MEDICAL LAW Charles Foster
MEDIEVAL BRITAIN John Gillingham and Ralph A. Griffiths
MEDIEVAL LITERATURE
Elaine Treharne
MEDIEVAL PHILOSOPHY
John Marenbon
MEMORY Jonathan K. Foster
METAPHYSICS Stephen Mumford
THE MEXICAN REVOLUTION
Alan Knight
MICHAEL FARADAY Frank A. J. L. James
MICROBIOLOGY Nicholas P. Money
MICROECONOMICS Avinash Dixit
MICROSCOPY Terence Allen
THE MIDDLE AGES Miri Rubin
MILITARY JUSTICE Eugene R. Fidell
MINERALS David Vaughan
MODERN ART David Cottington
MODERN CHINA Rana Mitter
MODERN DRAMA
Kirsten E. Shepherd-Barr
MODERN FRANCE Vanessa R. Schwartz
MODERN IRELAND Senia Pašeta
MODERN ITALY Anna Cento Bull
MODERN JAPAN
Christopher Goto-Jones
MODERN LATIN AMERICAN LITERATURE
Roberto González Echevarría
MODERN WAR Richard English
MODERNISM Christopher Butler
MOLECULAR BIOLOGY Aysha Divan and Janice A. Royds
MOLECULES Philip Ball
THE MONGOLS Morris Rossabi
MOONS David A. Rothery
MORMONISM
Richard Lyman Bushman
MOUNTAINS Martin F. Price
MUHAMMAD Jonathan A. C. Brown
MULTICULTURALISM Ali Rattansi
MUSIC Nicholas Cook
MYTH Robert A. Segal
THE NAPOLEONIC WARS
Mike Rapport
NATIONALISM Steven Grosby
NAVIGATION Jim Bennett
NELSON MANDELA Elleke Boehmer
NEOLIBERALISM Manfred Steger and Ravi Roy
NETWORKS Guido Caldarelli and Michele Catanzaro
THE NEW TESTAMENT
Luke Timothy Johnson
THE NEW TESTAMENT AS LITERATURE Kyle Keefer
NEWTON Robert Iliffe
NIETZSCHE Michael Tanner

NINETEENTH-CENTURY BRITAIN Christopher Harvie and H. C. G. Matthew
THE NORMAN CONQUEST George Garnett
NORTH AMERICAN INDIANS Theda Perdue and Michael D. Green
NORTHERN IRELAND Marc Mulholland
NOTHING Frank Close
NUCLEAR PHYSICS Frank Close
NUCLEAR POWER Maxwell Irvine
NUCLEAR WEAPONS Joseph M. Siracusa
NUMBERS Peter M. Higgins
NUTRITION David A. Bender
OBJECTIVITY Stephen Gaukroger
THE OLD TESTAMENT Michael D. Coogan
THE ORCHESTRA D. Kern Holoman
ORGANIC CHEMISTRY Graham Patrick
ORGANIZATIONS Mary Jo Hatch
PAGANISM Owen Davies
THE PALESTINIAN-ISRAELI CONFLICT Martin Bunton
PANDEMICS Christian W. McMillen
PARTICLE PHYSICS Frank Close
PAUL E. P. Sanders
PEACE Oliver P. Richmond
PENTECOSTALISM William K. Kay
THE PERIODIC TABLE Eric R. Scerri
PHILOSOPHY Edward Craig
PHILOSOPHY IN THE ISLAMIC WORLD Peter Adamson
PHILOSOPHY OF LAW Raymond Wacks
PHILOSOPHY OF SCIENCE Samir Okasha
PHOTOGRAPHY Steve Edwards
PHYSICAL CHEMISTRY Peter Atkins
PILGRIMAGE Ian Reader
PLAGUE Paul Slack
PLANETS David A. Rothery
PLANTS Timothy Walker
PLATE TECTONICS Peter Molnar
PLATO Julia Annas
POLITICAL PHILOSOPHY David Miller
POLITICS Kenneth Minogue
POPULISM Cas Mudde and Cristóbal Rovira Kaltwasser
POSTCOLONIALISM Robert Young
POSTMODERNISM Christopher Butler
POSTSTRUCTURALISM Catherine Belsey
PREHISTORY Chris Gosden
PRESOCRATIC PHILOSOPHY Catherine Osborne
PRIVACY Raymond Wacks
PROBABILITY John Haigh
PROGRESSIVISM Walter Nugent
PROTESTANTISM Mark A. Noll
PSYCHIATRY Tom Burns
PSYCHOANALYSIS Daniel Pick
PSYCHOLOGY Gillian Butler and Freda McManus
PSYCHOTHERAPY Tom Burns and Eva Burns-Lundgren
PUBLIC ADMINISTRATION Stella Z. Theodoulou and Ravi K. Roy
PUBLIC HEALTH Virginia Berridge
PURITANISM Francis J. Bremer
THE QUAKERS Pink Dandelion
QUANTUM THEORY John Polkinghorne
RACISM Ali Rattansi
RADIOACTIVITY Claudio Tuniz
RASTAFARI Ennis B. Edmonds
THE REAGAN REVOLUTION Gil Troy
REALITY Jan Westerhoff
THE REFORMATION Peter Marshall
RELATIVITY Russell Stannard
RELIGION IN AMERICA Timothy Beal
THE RENAISSANCE Jerry Brotton
RENAISSANCE ART Geraldine A. Johnson
REVOLUTIONS Jack A. Goldstone
RHETORIC Richard Toye
RISK Baruch Fischhoff and John Kadvany
RITUAL Barry Stephenson
RIVERS Nick Middleton
ROBOTICS Alan Winfield
ROCKS Jan Zalasiewicz
ROMAN BRITAIN Peter Salway
THE ROMAN EMPIRE Christopher Kelly
THE ROMAN REPUBLIC David M. Gwynn
ROMANTICISM Michael Ferber

ROUSSEAU Robert Wokler
RUSSELL A. C. Grayling
RUSSIAN HISTORY Geoffrey Hosking
RUSSIAN LITERATURE Catriona Kelly
THE RUSSIAN REVOLUTION S. A. Smith
SAVANNAS Peter A. Furley
SCHIZOPHRENIA Chris Frith and Eve Johnstone
SCHOPENHAUER Christopher Janaway
SCIENCE AND RELIGION Thomas Dixon
SCIENCE FICTION David Seed
THE SCIENTIFIC REVOLUTION Lawrence M. Principe
SCOTLAND Rab Houston
SEXUALITY Véronique Mottier
SHAKESPEARE'S COMEDIES Bart van Es
SIKHISM Eleanor Nesbitt
THE SILK ROAD James A. Millward
SLANG Jonathon Green
SLEEP Steven W. Lockley and Russell G. Foster
SOCIAL AND CULTURAL ANTHROPOLOGY John Monaghan and Peter Just
SOCIAL PSYCHOLOGY Richard J. Crisp
SOCIAL WORK Sally Holland and Jonathan Scourfield
SOCIALISM Michael Newman
SOCIOLINGUISTICS John Edwards
SOCIOLOGY Steve Bruce
SOCRATES C. C. W. Taylor
SOUND Mike Goldsmith
THE SOVIET UNION Stephen Lovell
THE SPANISH CIVIL WAR Helen Graham
SPANISH LITERATURE Jo Labanyi
SPINOZA Roger Scruton
SPIRITUALITY Philip Sheldrake
SPORT Mike Cronin
STARS Andrew King
STATISTICS David J. Hand
STEM CELLS Jonathan Slack
STRUCTURAL ENGINEERING David Blockley
STUART BRITAIN John Morrill
SUPERCONDUCTIVITY Stephen Blundell
SYMMETRY Ian Stewart
TAXATION Stephen Smith
TEETH Peter S. Ungar
TELESCOPES Geoff Cottrell
TERRORISM Charles Townshend
THEATRE Marvin Carlson
THEOLOGY David F. Ford
THOMAS AQUINAS Fergus Kerr
THOUGHT Tim Bayne
TIBETAN BUDDHISM Matthew T. Kapstein
TOCQUEVILLE Harvey C. Mansfield
TRAGEDY Adrian Poole
TRANSLATION Matthew Reynolds
THE TROJAN WAR Eric H. Cline
TRUST Katherine Hawley
THE TUDORS John Guy
TWENTIETH-CENTURY BRITAIN Kenneth O. Morgan
THE UNITED NATIONS Jussi M. Hanhimäki
THE U.S. CONGRESS Donald A. Ritchie
THE U.S. SUPREME COURT Linda Greenhouse
UTOPIANISM Lyman Tower Sargent
THE VIKINGS Julian Richards
VIRUSES Dorothy H. Crawford
VOLTAIRE Nicholas Cronk
WAR AND TECHNOLOGY Alex Roland
WATER John Finney
WEATHER Storm Dunlop
THE WELFARE STATE David Garland
WILLIAM SHAKESPEARE Stanley Wells
WITCHCRAFT Malcolm Gaskill
WITTGENSTEIN A. C. Grayling
WORK Stephen Fineman
WORLD MUSIC Philip Bohlman
THE WORLD TRADE ORGANIZATION Amrita Narlikar
WORLD WAR II Gerhard L. Weinberg
WRITING AND SCRIPT Andrew Robinson
ZIONISM Michael Stanislawski

Richard Passingham

COGNITIVE NEUROSCIENCE

A Very Short Introduction

OXFORD
UNIVERSITY PRESS

Great Clarendon Street, Oxford, OX2 6DP,
United Kingdom

Oxford University Press is a department of the University of Oxford.
It furthers the University's objective of excellence in research, scholarship,
and education by publishing worldwide. Oxford is a registered trade mark of
Oxford University Press in the UK and in certain other countries

First edition published in 2016
Impression: 7

Published in the United States of America by Oxford University Press
198 Madison Avenue, New York, NY 10016, United States of America

British Library Cataloguing in Publication Data
Data available

Library of Congress Control Number: 2016937496

ISBN 978–0–19–878622–1

Printed in Great Britain by
Ashford Colour Press Ltd, Gosport, Hampshire

To my undergraduate students,
who taught me how to explain things

Contents

Preface and acknowledgements

Cognitive neuroscience is a relatively new branch of science and this means that it is much less easy than in physics or chemistry to present an established view. So this book reflects the way I see things. I am grateful for comments by John Duncan, Eleanor Maguire, and James Rowe for keeping me on the straight and narrow.

I have assumed that, if we are to understand the results of experiments that use brain imaging, we need to know how information flows between brain areas. For this reason, I have included diagrams throughout the book that illustrate the anatomical connections between these areas. At first the reader may find these slightly forbidding, but though there are a profusion of terms there is nothing that is conceptually difficult.

I have also taken it to be important that the reader understands how the conclusions were reached. So I have often described the experiments in some detail and said who carried them out. The hope is that this will give some feeling for how the science is done and how it is progressing. To indicate that the argument is based on solid evidence, at the end I have included a list of the papers that provide the experimental data. There are many more than is usual for this series and many of them are technical. So I also give a separate list of recommended reading.

I have limited the scope of the book to the use of brain imaging to study human cognition. There has not been space to say what we have learned about possible mechanisms by recording the electrical activity of brain cells in other animals. It is a *very* short introduction.

It has become standard in the brain imaging literature to use the word 'volunteers' for those who take part in an experiment. I have refused to do so because it is really quite irrelevant that they volunteered rather than being dragged by the hair into the scanner. It is more relevant that they are people and so that is the word that I prefer to use.

I am aware that the account I give is greatly oversimplified. The reader may forgive me but my professional colleagues will probably slay me.

List of illustrations

Chapter 1
A recent field

When I learned psychology as a student at Oxford we were not encouraged to ask questions about the mind. It was in the early 1960s and psychology was still strongly influenced by the doctrine of behaviourism. This maintained that science could measure stimuli (inputs) and responses (outputs) but that it was unscientific to speculate about what happened in-between. The reason was that this happened in the head and so there was no objective way of finding out. So for us, psychology was mainly about the behaviour of rats and pigeons because it was easy to control the stimuli and measure the responses. No wonder we found the subject terribly dull.

At the same time we had to go to lectures in philosophy. There were philosophers who were happy to talk about the mind, but many of them believed that the mind was separate from the brain. This philosophical position is termed 'dualism'. Unfortunately these philosophers were no better able to suggest how the mind and brain interact than Descartes had been nearly 400 years ago. So there was little excitement to be had in hearing of such a lack of progress.

The only novelty was provided by the lectures by Gilbert Ryle. He argued that the dualism was a 'bad mistake' because it invoked a 'ghost in the machine'. But it was not clear, when the ghost had been banished, what was left for us to say about our rats or pigeons.

Cognitive psychology

So what, if anything, has changed? Even at that time behaviourism was beginning to crumble. The reason is that objective ways were being found of settling what *must* be happening in the head. As students we learned, for example, of the experiments by Donald Broadbent and Anne Treisman in which subjects listened on headphones while a series of numbers or words was played to the right ear and another series to the left. The people in the experiments were instructed to attend to the items they heard in one ear and the finding was that they were unable to remember those that they heard in the other. So Donald Broadbent argued the brain must filter out the items arriving from the unattended ear. It was experiments of this sort, together with ones on perception by Ulrig Neisser and on memory by George Miller, that led to the founding of the science of 'cognitive psychology'.

One way of showing what must be happening in the head is to produce a diagram to illustrate the flow of information in the system. Donald Broadbent pioneered the use of diagrams of this sort and he included a filter in his model so as to account for the empirical results on attention. These models were called 'black box' diagrams because they were drawn as a set of boxes linked by arrows; the boxes were said to be 'black' because at the time it was not known how the components worked or where they were to be found in the brain.

Attempts can be made to find out by studying the effects of damage to the brain in patients. If the damage is relatively restricted it should be possible to make suggestions as to the functions of the missing components. And by the late 1970s enough was established in this way to encourage the development of a new science. While chatting in a taxi together, Michael

Gazzaniga and George Miller came up with a name for it, and this was 'cognitive neuroscience'.

Cognitive neuroscience

They could not have conceived that within a decade psychologists would have access to techniques that allow us to visualize the activity of the brain in healthy people, and to do so while they carry out psychological tasks. These methods include positron emission tomography (PET), developed in the 1980s, and functional magnetic brain imaging (fMRI), developed in the 1990s. These techniques have transformed the field of cognitive neuroscience. There have now been nearly 30,000 experiments conducted using fMRI alone. Figure 1 shows one of the most popular scanners used for fMRI.

And so psychology has altered out of all recognition from the subject that I learned as a student. As lay people had always

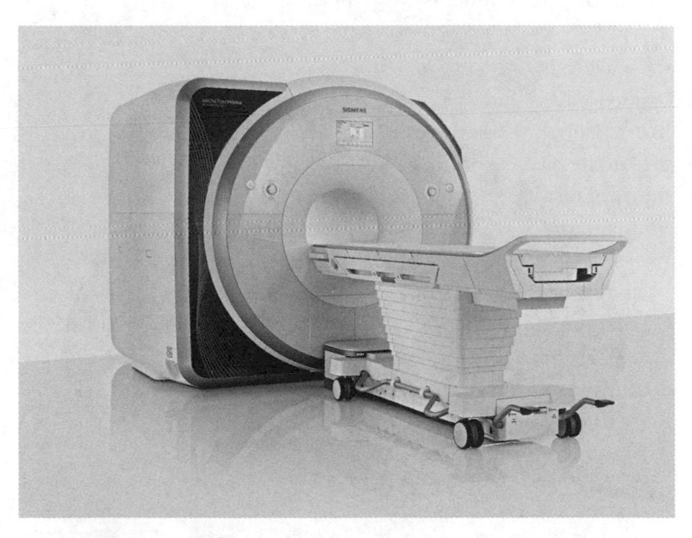

1. A popular type of functional magnetic resonance scanner.

supposed, it is indeed the science of the mind and in particular the human mind. And since many psychology departments now have access to brain scanners, it is common to see a line of posters along the corridors that show images of the human brain at work.

The interpretation of brain images

These images are both beautiful and alluring (Figure 2); so it is no wonder that they excite the interest of laypeople and the media. We read, for example, that the trials of the teen years are due to the impoverishment of their frontal lobes or that brain imaging has proved that women are unable to think logically. Such claims make good headlines.

Science reporters who pass on claims such as these can show their ignorance in one of three ways. The first is forgivable, the second suggests that they do not understand science, and the third indicates a wider difficulty with logic.

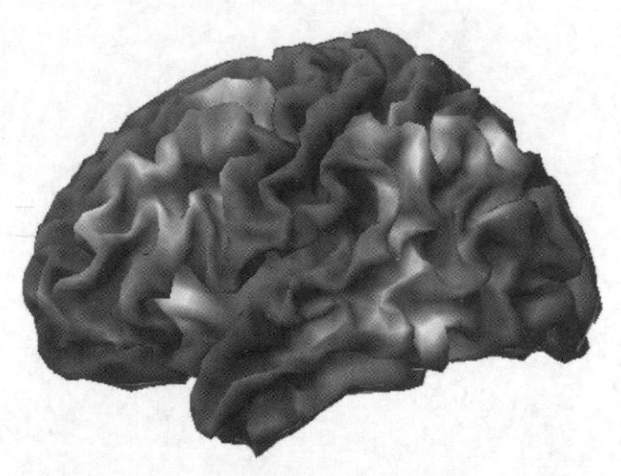

2. Typical map of activations as visualized with fMRI.

The first misunderstanding is to suppose that the coloured patches in the brain images show where there is an increase in the activity of brain cells or neurons. In fact they are only an indirect measure of brain activity, since the signal to which PET and fMRI are sensitive reflects changes in the blood supply; in the case of fMRI this is the ratio of oxygenated to de-oxygenated blood. The rationale is that when neurons increase their activity there is an increase in the supply of arterial blood to that region, bringing the oxygen and glucose that are needed for metabolism. It is because the measure is indirect that in the imaging literature there is said to be 'activation' in an area rather than 'activity'.

The second misunderstanding is to believe that the images show all the areas in which there was activation during the experiment. They do not. Instead, as in science in general, the images result from a comparison between experimental and control conditions.

The logic is the same as in medicine. When doctors test whether a new medication works they compare it with a dummy pill or placebo. The reason is that the patients might get better simply because they know that they have been given a pill by someone in a white coat and half-rimmed glasses. So the test is whether the patients who take the genuine pill do better than those who take the dummy pill. It is the *difference* that matters.

The same method is used in brain imaging. It can be conveniently illustrated by taking an experiment that is described more fully in Chapter 4. The aim was to find out where there is activation when people have to think about the way in which objects are related. So during scanning the people were shown pictures like the one in Figure 3a; here they had to say whether the spanner or the saw is the one that is most obviously associated with the pliers. The answer is the spanner because both can be used to grip.

But if people were simply scanned while they did this task, the images would show the activations for viewing the pictures and

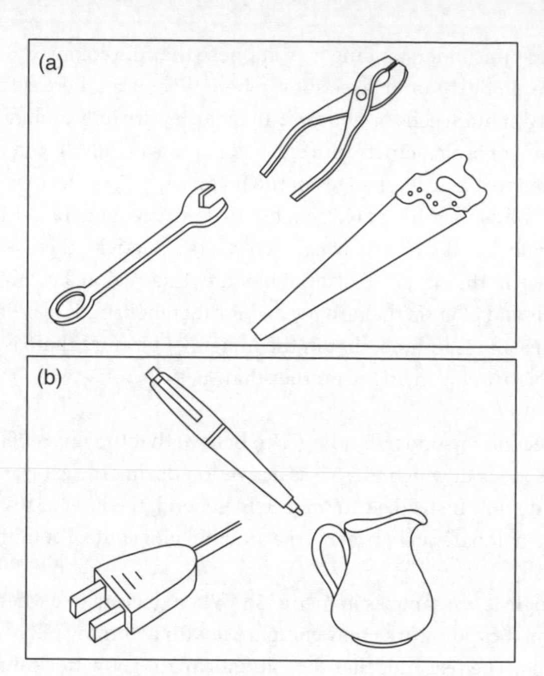

3. Pictures in (a) experimental and (b) control conditions for experiment on semantic knowledge.

for making the choice. Yet the experiment was designed to investigate the ability to retrieve knowledge about how the objects were used. So the people were also scanned in a comparison or 'control' condition (Figure 3b). For this they had to decide which of the two objects was the same size in life as the one above. Here too they viewed pictures and made a choice, but they did not have to consider how the objects are used.

So the analysis of the data compared the activations in the experimental association condition (E) (Figure 3a) with the activations in the control size condition (C) (Figure 3b) and calculated the *difference* between them (E – C). The difference image (E – C) removed or subtracted out the activations for viewing pictures and responding. So the resulting image only showed the activations that were specific to knowledge of the use of tools. Most of the results reported in this book come from difference images of this sort.

However it is the third misunderstanding that is fundamental. Science is interested in causes and it is easy to suppose that the pattern of activation in the brain must be the *cause* of the mental state or behaviour. In one sense, of course, it is: without the brain there would be no mental state.

But there is another sense in which the pattern of activation need not provide an explanation of the mental state. This can be illustrated by taking as an example the images produced when patients are scanned while they are depressed. The standard finding is that there are two areas in which there is a difference in the degree of activation in the patients compared with people who are not depressed. One area is called the amygdala (Figure 4) and the other the subgenual cingulate cortex (Figure 4). Beware the science reporter who tells you that this image proves that depression is a brain disorder.

Of course the brain is involved, but the problem is that the term 'brain disorder' implies some defect in its internal workings. But the same pattern of activation is found if people who are not depressed are simply instructed to think sad thoughts while in the scanner. And their brain is not disordered. Thus, the image could simply be telling us what we know already, namely that people who are depressed are thinking sad thoughts.

So the cause of the depressed state could be internal, perhaps due to the inheritance of particular genes. Or it could be external, perhaps due to the breakup of the marriage. The scans alone will not tell us.

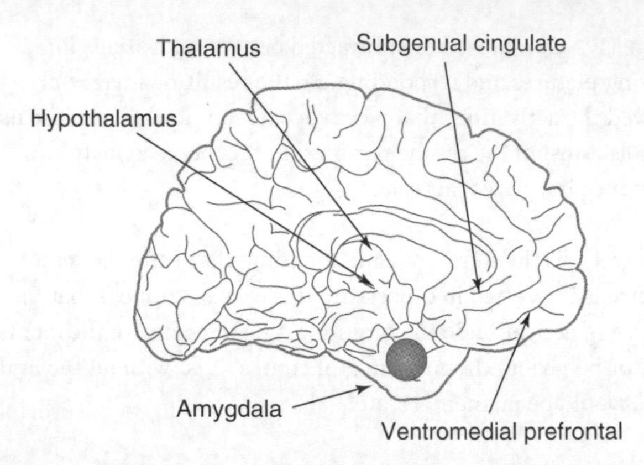

4. Medial or inner surface of the left hemisphere of the human brain. The diagram shows the areas that are activated during studies on depression, sexual attraction, and preference for juice. The amygdala lies under the surface of the cortex.

This point is one that many in the media have failed to appreciate. And it matters. Take studies in which gay men are scanned while they are viewing erotic pictures of men; when they do so there is activation in a structure called the hypothalamus (Figure 4). It is easy to jump to the conclusion that the men are gay because of some abnormality in their brain. But the findings are the same when straight men view erotic pictures of women. The images simply tell us which gender the individuals find attractive. They tell us nothing about why some men are gay.

If you are not convinced, then take the Pepsi challenge. If people who prefer Pepsi are scanned, there is more activation in the ventromedial prefrontal cortex (Figure 4) when they drink Pepsi rather than Coca-Cola. Naively you might think that this shows that a preference for Pepsi is due to some kink in the brain. But if people who prefer Coca-Cola are scanned, there is more activation in the same area when they drink Coca-Cola. So all the scans actually indicate is which drink the person likes; they do not tell us why. Perhaps all that expensive advertising pays off.

The conclusion is that what brain images show us is the state of the brain when people are in particular mental states. They do not *necessarily* tell us why they are in those states. It could be because of past history or current events in the world. Brain explanations do not necessarily trump environmental explanations.

The reason why this matters is that the aim of psychology is to help us to understand people; and people are not simply brains. Like other animals, they have a body which grows up and ages, and they live in a physical and social environment. So the province of psychology cannot simply be reduced to the province of neurology. There are facts about people that are not just facts about their brains.

Answering psychological questions

It follows that there are many questions in psychology for which the answers must be sought by carrying out psychological experiments. Is phonics the best way of teaching children to read? Is it true that the capacity of our short-term memory is limited to around seven items? Why do people hold social stereotypes concerning out-groups? And so on.

Questions such as these will not be answered by simply scanning people and noting that there is activation in area X when they are in mental state A. There *has* to be some change in the brain, and the danger is that all the experiment tells us is where that is. Studies that simply do this may tell us about the brain, but they do not tell us about psychology.

There are, however, three issues that imaging can address. The first is how to account for human capacities. For example, people differ in their intelligence, and tests of intelligence typically test a range of abilities; yet the different subtests tend to give similar results. Why should this be so?

The second issue concerns human limitations. We know, for example, that multitasking can lead to inefficient performance, but it is not enough to say that there is a limit to our capacity to attend to two things at once. Why is the system not designed to handle many messages at the same time?

The final way in which imaging can advance psychology is by providing an explanation for the psychological effects of disorders of the nervous system. Many amputees continue to feel sensation in their non-existent limb, and some also feel chronic pain there. How can that be explained?

What questions such as this have in common is that it is not adequate to answer them simply by talking in terms of past learning or social interactions. The phenomena point to what the human mind can and cannot do and what happens when it goes wrong, whether from illness, damage, or age.

So this book takes up the challenge. It seeks explanations. Each chapter considers the sorts of questions that a layperson or a psychologist might ask, and tries to show how understanding the activity of the human brain can help us to provide answers. It is true that in some cases these answers confirm what psychologists have suspected. But it would be sad indeed if psychologists had everything wrong, and science always looks for confirmation. And there are other cases in which the answers are genuinely novel.

Each chapter starts with three questions. The discussion that follows then provides the general background and it is on this basis that the answers are given in a box at the end. These are stated as facts but the reader should be warned that in science facts can and do change.

Chapter 2
Perceiving

Questions

1. Do you need to recognize an object to know how to handle it?
2. Why do some amputees continue to feel their arm even though it is absent?
3. Why do some people see colours when they read or hear words?

Suppose I put a pencil on the table and ask you to write your name. Common sense tells us that you need to be able to perceive the way in which the pencil is oriented in order to know how to pick it up properly. It is one of the delights of science that common sense is so often wrong.

In this case we know this to be true because David Milner and Mel Goodale have described a patient called Dee who is unable to identify the orientation, yet she is able to pick the pencil up straight away. Dee came to their notice in the clinic because of her inability to recognize objects; this was the result of damage to her brain from carbon monoxide poisoning when a water heater leaked while she was on holiday.

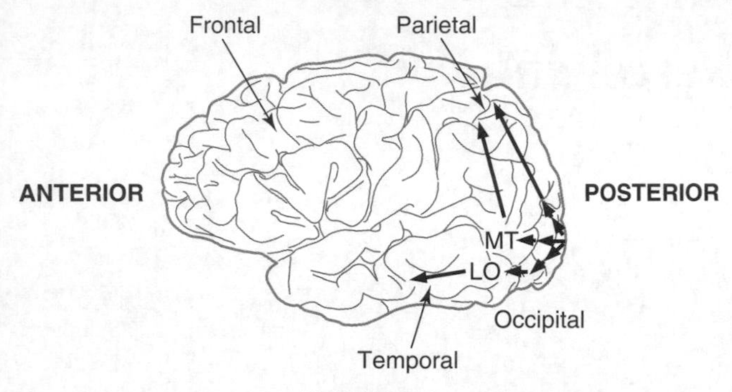

5. Connections of the ventral visual pathway are shown as thick arrows from primary visual cortex V1 via the lateral occipital complex (LO) to the temporal lobe. Connections of the dorsal visual pathway are shown as thick arrows from primary visual cortex V1 via the middle temporal complex (MT) to the parietal lobe and also from V1 via other intermediate areas to the parietal lobe. Dorsal = top surface of the brain, ventral = bottom surface of the brain, anterior = front of brain, posterior = back of brain.

One way of explaining what Dee can and cannot do is to suppose that there must be two systems that are relatively independent, one that is specialized for the perception of shape and another that is specialized for using shape to guide action. The suggestion is that in Dee's brain the first must have been damaged whereas the second must be intact. Fortunately, there is a way to find out and this is to scan her brain.

The results show that the damage in Dee's brain is in an area that is activated when other people recognize objects. This area is referred to as the lateral occipital or LO complex; it lies at the border between the occipital and temporal lobe (Figure 5). There is no signal here when Dee views objects and this accounts for her failure to recognize objects. So the LO complex is critical for the ability to use the shape of an object to identify it.

The picture is quite different when Dee is required to reach for and grasp an object in the scanner. Now activation can be detected in an area that lies in the parietal lobe (Figure 5), and the same area is activated when healthy people are scanned. This area is critical for the ability to use the shape and orientation of an object to guide the way in which the hand approaches it.

The reader may be worried that these conclusions are based on the study of a single patient. But we cannot deliberately cause selective damage to an area such as the LO complex, and so we should welcome the fact that occasionally nature obliges. Science often benefits from occasional accidents, as when Alexander Fleming noticed the effect of penicillin mould in his petri dish. It is to the credit of David Milner and Mel Goodale that they saw the significance of their observation.

Parallel pathways

The lesson to be learned from Dee's performance is that the brain is not a single pathway from input to output but that there are separate pathways and that pathways also diverge. Information from vision, hearing, touch, and smell is relayed from the sense organs to separate regions of the brain (Figure 6). These are referred to as the primary sensory areas. The information is then relayed through a series of secondary areas.

It is the convention to give each of the primary sensory areas the number 1, as in V1 for the primary visual area, and to number the secondary areas in ascending order. In the case of vision these are V2, V3, V4, and so on. The primary visual cortex relays information via these and other areas to the temporal lobe and the parietal lobe (Figure 5).

These relays are termed the ventral and dorsal visual pathways. As can be seen from Figure 5, the term 'ventral' means towards the lower surface and the term 'dorsal' towards the upper surface of

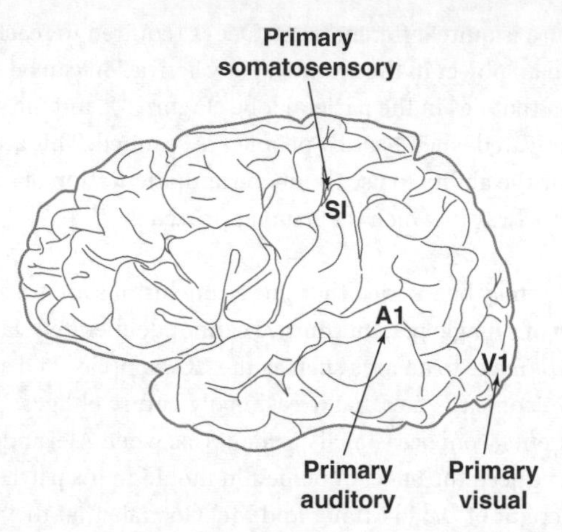

6. Primary sensory areas. V1 = primary visual, A1 = primary auditory, S1 = primary somatosensory.

the primate brain. Anatomy is an ancient discipline and it still retains some arcane terminology.

Apart from differing in their location the ventral and dorsal visual systems also differ in their outputs. Only the dorsal one has direct connections with the areas in the frontal lobe that control movement. This is why it is the dorsal system that is specialized for using vision to guide action, as in the case of picking up a pencil.

The dorsal pathway also carries information about the movement of objects. This is analysed in the human middle temporal or MT complex (Figure 5) and there are then connections from the MT complex to the parietal cortex. So, given that the parietal cortex receives information about movement of an object, this can be used to guide the hand or eyes (Figure 5).

The fact that there are separate or 'parallel' pathways means that different operations are carried out simultaneously: we see at the same time as we hear. So the brain is unlike a desktop computer in which the instructions are followed one after another in a series. Instead the brain has a parallel architecture, meaning that many different operations can be carried out at the same time. The overwhelming advantage is in the speed of processing.

It is this organization that also explains how it is possible for damage to the brain to impair one ability while another is preserved. For example, the neurologist Joseph Zihl has described patient LM who can see a car on the road ahead but is unable to perceive its motion. Instead she sees a series of separate images, as if a film had been slowed down until it was a succession of stills. The explanation is that she has a lesion that includes the human MT complex. But the lesion does not extend down to the LO complex, and that is why she is able to recognize the car by its shape.

Another way of showing that there are separate mechanisms for the perception of motion, shape, and colour is to use brain imaging. Semir Zeki pioneered the use of imaging to do this. For example, he has scanned people while they viewed displays of dots. When the dots were moving, the signal was in the MT complex. When the dots were coloured, the signal was in an area that, to the despair of the reader, some refer to as V4 and others as V8.

Areas and their connections

It is observations such as these that demonstrate that different brain areas perform different operations. The brain is not an amorphous mass like a lump of porridge but a patchwork of discrete areas. These can be distinguished by examining the tissue with a microscope. Each area consists of a host of neurons arranged in six layers, each layer being many neurons thick. The areas differ in the thickness of the layers, the density of the packing, and the frequency of the different types of neuron.

Each neuron has a cell body and a long 'axon' or fibre. These fibres are collected into bundles and they can travel long distances so as to connect areas that are far apart. It is the pattern of these connections that means that different areas can perform different operations. This is the principle of 'localization of function', and it can hold because each area receives a unique pattern of connections and sends out a unique pattern of connections to other areas. The incoming connections determine the information that the area can process; the outgoing connections determine the influence that it can exert on other areas.

Mapping

There is another organizing principle. This is that the outside world is mapped onto the brain in a point-to-point fashion. For example, in the primary somatosensory area (S1) (Figure 6) there is a map of the body with different parts responding when the hands, body, or legs are touched. Similarly, in the primary visual cortex (V1) (Figure 6) there is a 'retinotopic' map. What this means is that the retina is mapped onto the cortex such that the location of the activity in the cortex depends on which part of the retina is stimulated by light.

However, these maps are distorted. In the case of the somatic map the distortion relates not to the size of the different parts of the body but to their relative importance. For example, because we use our hands for manipulating objects, much more tissue is devoted to the hand than to the foot. In the case of the primary visual cortex, more tissue is devoted to central than peripheral vision. In the centre of the eye there is a pit or fovea and it is here that there is the greatest density of light receptors. When we inspect an object it is with the fovea that we focus our gaze.

One way of demonstrating the map in the primary visual cortex is to stimulate the neurons artificially. Alan Cowey and Vincent Walsh have done this by imposing magnetic pulses by means of a metal coil on the scalp, a technique called 'transcranial magnetic brain stimulation' (TMS). The method may appear heroic, but as far as we know it is harmless. The effect of stimulating the visual cortex is that observers report seeing flashes of light or 'phosphenes' and the apparent location of the phosphenes differs depending on where the stimulation is applied on the scalp. Some people see these phosphenes when they have a migraine.

Phosphenes can be evoked even if the visual cortex is stimulated in people who have been totally blind from birth due to damage to the retina. So, even if the peripheral sense organ is damaged or missing, activity in the sensory cortex is perceived as originating from the outside world.

This helps to explain why some amputees continue to feel sensations from their limb even though it is absent. Depending on where the stump is touched, forearm amputees report feeling that different fingers are being touched. When these amputees are scanned, it turns out that touch on the stump leads to activation in the hand area of their somatosensory cortex; and furthermore all the fingers are represented there just as would be expected if the hand was intact. The implication is that in amputees, if there is activity in the hand area, it is felt as if it originates in the absent hand or arm. Hence the illusion.

Putting the senses together

Given that different aspects of our sensory world are analysed in different brain regions, we need to explain how it is that we perceive the world as a whole. It is no good seeing an object if the sight of it gives you no idea how big it will feel when you pick it up.

There are two ways that sensory integration is achieved. Integration within any one sensory modality is achieved by connections within that system. For example, in vision these connect areas in which the neurons respond to colour and areas in which the neurons respond to shape.

Integration between sensory modalities is achieved by connections to common areas, and these are referred to as 'multimodal' areas. They include the parietal cortex (Figure 5) and the prefrontal cortex (Figure 16, p. 67). The evidence that these areas are multimodal is that they can be activated irrespective of whether the stimuli are visual, auditory, or tactile.

Fortunately we have a way of demonstrating the relevant connections in the human brain. This is called 'diffusion weighted imaging' (DWI). It makes use of the fact that the scanner can be sensitized to the movement of water in the brain; it is possible to chart the connections in this way because water diffuses along the fibre tracts. So we can visualize the orientation of these tracts and also estimate their points of origin and termination.

If integration between the sense modalities depends on the connections, it should be possible to show that the connections are abnormal in people who show unusually strong integration between the senses. There are people with synaesthesia who see colours when they hear speech or read words. In one study, synaesthetes reported seeing colours when they heard words and when they were scanned there was a corresponding activation in the visual area V4/V8—the area in which there is a signal when people are actually shown colours.

The study confirms that the synaesthetes are seeing what they claim to see, but it does not explain why they are doing so. The answer is suggested by a study that used diffusion weighted imaging to compare the connectivity of the brains of people who did or did not have synaesthesia. The connections within the parietal cortex, one of the multimodal areas, were stronger in the synaesthetes. The same study also examined in particular those synaesthetes who saw colours when they read; and in this case the connections within the inferior temporal cortex were stronger in the synaesthetes. The suggestion is that it is these connections that link the visual areas that analyse shape and colour.

Recognizing objects

Objects are recognized on the basis of their shape and colour and it has already been established that the LO complex is necessary for object recognition. But this does not tell us how that recognition occurs. To understand this we need to appreciate that processing occurs via successive stages in the visual relay (Figure 5).

These stages differ in that the earlier ones represent small elements of the visual scene whereas later ones have a more complete representation. We know this because Kalanit Grill-Spector and Rafael Malach have scanned people while they viewed a series of pictures that differed in how complete they were. Some showed objects that were complete, some two halves, some four quarters, and so on, with differing degrees of scrambling. In the earliest areas V1–V3 the activation was maximal for the scrambled pictures, in V4 for ones that were less scrambled and in the LO complex for pictures that were either complete or in halves.

So the earlier stages process the elements and the later stages integrate them. It is thought that this is achieved via a hierarchical arrangement in which each higher-order neuron in area B receives an input from many lower-order neurons in area A, each neuron in area C receives an input from many neurons in area B, and so on through successive stages.

An arrangement of this sort would also allow us to learn that an object is the same even though viewed from different angles. As we approach an object in everyday life we see it from different viewpoints at different times. At one time the object might be to the left of our gaze and at another to the right. At one time the object might appear small because it is far away and at another large because we are nearer to it. And at one time we might see the object from one angle and at another from the opposite angle. Yet we need to appreciate that the object is the same in all cases.

Human beings are very good at solving this task. Given a long childhood we have plenty of time to learn to recognize and identify the objects in our world. By moving towards objects or moving around them, we can learn that the object is one and the same even though seen from different views. The evidence that it is the same object is that the different views occur within seconds of each other.

We believe that the visual system is organized in a way so as to make this learning possible. Lower-order neurons respond maximally when we see an object from a particular view whereas higher-order neurons integrate the information from lower-order neurons and are thus able to learn to respond irrespective of the view. The suggestion is that it is in this way that we form a 'view-independent' representation of the object.

There is evidence that is consistent with this proposal. It comes from a brain imaging experiment that made use of the fact that if the same stimulus is presented repeatedly within a short period there is less activation with repetition. This effect is referred to as 'adaptive suppression'. The reason that it occurs is that the system is tuned to treat novel objects as being of more interest than ones with which we are familiar. Thus, the method can be used to find out where there is invariance in the visual system: if the object is classified as being the same on different occasions, even though seen from different views, there will be evidence of adaptive suppression.

So Kalanit Grill-Spector and Rafael Malach scanned people while presenting the same object repeatedly, varying the location, size, or viewpoint. There was little adaptation in earlier visual regions, but an adaptive suppression effect could be demonstrated in the more anterior part of the LO complex. This supports the notion that higher-order areas form representations that are invariant by integrating information from lower-order areas.

Classifying objects

It is not enough simply to recognize what we see; we also have to make sense of it. This involves classifying things, whether animate or inanimate. For example we categorize animals as primates, birds, insects, and so on. So Jim Haxby has presented pictures of animals and asked people to classify them. The same people were then scanned while they viewed them, and this made it possible to relate the judgements they made to the patterns of activation as found in the visual system.

The images were analysed using a method that is called a 'multi-voxel pattern analysis'. The brain image is made up of very many 'voxels' or three-dimensional pixels or picture elements. So instead of simply measuring the degree of activation in an area, we can also analyse the pattern of activation *across* voxels to see if it differs depending on what the people are looking at.

When the analysis was performed, the results differed according to the stage of the visual system. It was not possible to distinguish the different categories by analysing the pattern of the activation in the primary visual cortex V1. But in higher visual areas in the inferior temporal cortex there was a relation between the patterns of activation for the different animals and the way in which the people had classified the animals.

This classification has, of course, been learned during childhood and adolescence. As a child we are taught that different animals are primates or birds. The suggestion is that the representation of these categories is learned by groups of neurons in higher areas through the association of inputs from lower-order areas. As is the case with object recognition, this depends on a hierarchical organization.

Perceptual awareness

It is this organization that explains why, if magnetic pulses are applied to the earliest stages of visual processing, people report that they only see elementary flashes of light. But how can we explain the fact that they are *aware* of these flashes?

This issue was long thought to be a topic for philosophy rather than science. Then Larry Weiskrantz made an astonishing discovery. He was studying two patients who had damage to their primary visual cortex in one hemisphere. As expected, they were blind in one half of their visual field, meaning that they were unable to see anything that was presented on the side of space that was opposite to their lesion.

But surprisingly they could still *guess* accurately the orientation of a line or the direction in which it was moving. And this was true even though they were not aware of seeing the line. The explanation for this 'blindsight', as he called it, is that visual information can still reach the cortex through routes that bypass the primary visual cortex.

This demonstrates that the primary visual cortex is *necessary* for visual awareness, but it does not tell us whether it is *sufficient*. To find out we can scan people when they are or are not visually aware and see what, if any, other areas are activated.

One way of doing this is to make use of a technique called 'backward masking'. If a stimulus, say a letter, is followed in 20 milliseconds by a pattern in the same location, most people are unable to report what the letter was; but if the pattern follows in 100 milliseconds many people can do so. This technique means that we can compare the activations when subjects are aware with the activations when they are not. It is a common finding that when subjects *can* report the letter the activations extend outside the ventral visual system to the parietal (Figure 5) and dorsal prefrontal cortex (Figure 16, p. 67), and that the activations in these two areas vary in synchrony.

But the philosopher Ned Block has suggested that there is a problem in interpreting these findings. It could be that these activations do not relate to visual awareness itself, to 'phenomenal awareness'. They might instead relate to the requirement to report that you are aware. He suggests that this requires 'access' to awareness.

There is a way of evading this problem and this is not to ask the person whether they are aware or not. A very crude way of doing this is simply to compare the situations where the person lies in the scanner with their eyes open or closed. In the first case they are visually aware, in the second they are not. One study found that when people had their eyes open there was activation in the visual areas V1, V2, and V4. However, there was no *additional* activation in the parietal or prefrontal cortex as compared with the situation in which the people lay with their eyes closed. The implication is that it is activation in the ventral visual stream that underlies *phenomenal* awareness for vision. If the parietal and prefrontal cortex contributed, we would have expected there to be an increase in their activation when the people opened their eyes.

A more sophisticated approach is to study synaesthetes since they are phenomenally aware of seeing colour even when no colour is present. A study has already been mentioned in which there was activation in area V4/V8 when they had the illusion of seeing colour. However, again there was no *additional* activation in the parietal or prefrontal cortex compared with the scans of non-synaesthetes.

There were, however, two other areas that were activated in both studies. One was the anterior part of the insula (Figure 7, p. 29), the other the anterior cingulate cortex (Figure 20, p. 85). These areas are interconnected and they are known to respond when a salient event is detected. This has led to the suggestion that they form a 'salience network'.

There is yet another way in which we can study awareness and this is to manipulate it by giving an anaesthetic such as propofol. So Irene Tracey and her colleagues scanned people while they were being anaesthetized. The degree of consciousness was assessed by speaking words aloud or by applying a painful stimulus. As expected, there were decreases in the activation in the insula and anterior cingulate cortex as the anaesthetic took hold. However, there were also decreases in the thalamus (Figure 4, p. 8), a structure that relays sensory information to the cortex. And as a result, the auditory cortex became unresponsive to words and the somatosensory cortex unresponsive to painful stimuli.

The philosopher David Chalmers would object that none of these results explain what accounts for the state of awareness; he refers to this as 'the Hard Problem'. But the problem can be simplified. The issue is usually discussed in relation to vision, with questions such as why we see the colour red as red or how it is that we view the world as if it were a film out there.

But now consider pain or the 'ouch' response and the problem does not seem quite so intractable. A dog whimpers after treading on a spine; and by analogy with our own experience we say that the dog 'feels' the pain. We take the analogy to be justified because dogs are mammals, have similar pain pathways, and fail to respond to the same painful stimulus when anaesthetized.

Of course, even if it were agreed that the dog feels pain, the sceptic could still argue that it lacks the ability to tell when it is in pain. But this is an issue that is open to testing. The dog could be trained to press one lever when in pain, and another when not in pain. Success on this test would mean that the dog can discriminate between the two states. It would not be possible to explain away the results by claiming that the dog is simply wired to make a reflex response.

Thus the pain system provides a simple model for the study of awareness. An initial line of attack would be to investigate further the functions of several structures. Of these the thalamus is the most important because, beyond knowing that it serves as a gateway to the cortex, we know too little about what it does. Of the cortical areas, the ones that deserve further study are the anterior insula and the anterior cingulate cortex with which it is connected. Both respond to pain but the use of the term 'salience network' is an admission of failure. Hence the need for further research.

Awareness is just one of the issues that were once a matter of speculation amongst philosophers but have now become tractable for science. It is easy to forget this. It is only a little over a century ago that the French philosopher Henri Bergson argued that it was necessary to invoke an *élan vital* or vital force to explain the development of organisms. But now we know that control genes direct the course of development. When we have a full account of how the brain receives and processes sensory information it is unlikely that we will need to invoke a mysterious emanation to explain awareness.

Answers

1. Object recognition depends on the ventral visual system in which both the recognition of objects and their classification is achieved via a hierarchical analysis. This system for identifying an object is separate from the dorsal visual system which is involved in guiding the hand so as to grasp the object. Even if one system is damaged, the other can still

operate. Demonstrations such as this are evidence for the localization of function in the brain.

2. The explanation of why amputees continue to feel the absent arm is that there is still a representation of the forearm and hand in their somatosensory cortex. Stimulation of the stump can evoke cortical activity. Just as stimulation of the visual cortex evokes phosphenes that appear to come from outside, so activity in the somatosensory cortex is perceived as touch to the forearm even though it is missing.
3. Information about the senses is integrated via connections within and between sensory systems. In synaesthesia the connections between the areas for analysing shape and colour are abnormally strong. In synaesthetes the illusion of seeing colour is related to activation in the visual area V4/V8. More generally, it appears that activity in early sensory areas together with activity in the insula and anterior cingulate cortex is sufficient for phenomenal awareness.

Chapter 3
Attending

Questions

1. Why do some patients neglect things to their left after a stroke?
2. Why does attending to something else decrease the feeling of pain?
3. Why is it dangerous to use a mobile phone while driving?

Awareness can be selective. Suppose that you are listening to music through headphones and trying to solve a Sudoku puzzle at the same time. You realize after a while that you have missed a whole passage of the music. The way you account for this is to say that you 'were not attending' to it.

But it is not obvious why there should be a limit to what you are aware of. After all, the information that comes from your various senses is processed in different areas of the brain. And as mentioned in Chapter 2 the processing occurs in parallel, meaning that what you see is analysed at the same time as what you hear or touch. Given that there are an estimated 100 billion neurons in the human brain it seems unlikely that there is a limit to its processing resources.

So there must be some other explanation for why you failed to hear the music. The answer is that only some of the incoming sensory information is relevant for solving the Sudoku puzzle. The irrelevant information only acts to distract you from performing the task efficiently. So it is not that there is a limit to the capacity of the system but that at any one time you are only interested in some of the information that is available.

This is particularly true for what we see. Our visual environment is very rich, full of objects of various shapes and sizes embedded in a complex background scene. So we scan the world, and what we look for and attend to depends on what task we are doing at the time.

Looking and attending

In the laboratory it is the experimenter who specifies the task. Michael Posner came up with a classic paradigm in which the observers are instructed to look out for the appearance of a visual target in the periphery of their vision. However, they are also to keep their eyes fixed on a central dot. So their attention is said to be 'covert'; in other words there should be no overt eye movements. On 80 per cent of trials the target appears on the side to which the people are attending, but on 20 per cent of trials it appears on the unattended side. This arrangement means that it is possible to demonstrate the benefit of attending: the targets on the attended side are detected more quickly than the targets on the unattended side.

If people are scanned while they are covertly attending to one side, the effect is an enhancement in the activation; this means the activation is greater when attending than when not doing so. The enhancement is found in two regions. One is the cortex that lies in the intraparietal sulcus, a deep fissure in the parietal cortex (Figure 7). The other is the area of the prefrontal cortex referred to as the frontal eye field (Figure 7).

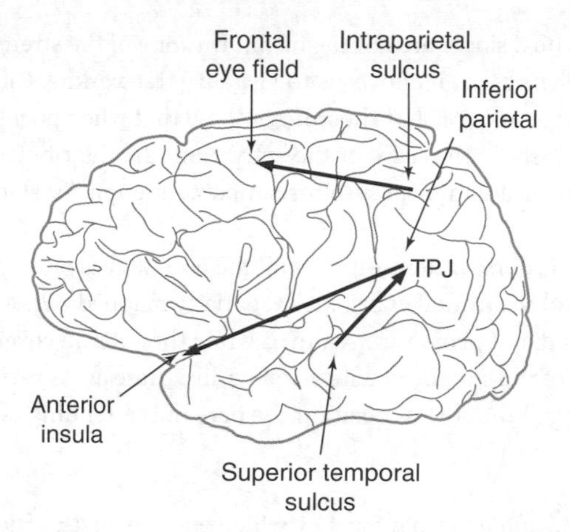

7. Dorsal and ventral attention systems. TPJ = temporo-parietal junction.

Both areas are also engaged when people actually move their eyes or simply plan to move their eyes. So when you attend to the periphery while gazing centrally, the enhanced activation reflects the fact that you are preparing to move your eyes. And in fact it is so tempting to actually move them that the experimenter needs to measure your eye movements to make sure that you don't peek.

The two areas of the brain involved, the cortex in the intraparietal sulcus and the frontal eye field, are part of what is called the 'dorsal attention system' (Figure 7). The term 'anatomical system' is used for a group of areas that are closely interconnected. As a result of these connections the spontaneous activity in these areas varies in synchrony. This is called 'resting state covariance' because the measurements are taken when the people are simply resting in the scanner.

The concept of a system is fundamental in cognitive neuroscience. Though areas differ in the operations that they perform, they do not operate in isolation. To put it bluntly, no behavioural task depends on a single area of the brain. It is one of the strengths of brain imaging that it allows us to visualize the workings of the whole brain at once, and the images show that when people perform tasks in the scanner the activations are distributed across the brain. So anatomical systems support functional systems.

Within the dorsal attention system, the cortex in the intraparietal sulcus and the frontal eye fields is spatially mapped. What this means is that if people are scanned while they attend covertly, the location of the enhanced activation within these areas varies according to the spatial position the person is attending to in peripheral vision.

Thus, one of the mechanisms by which we prioritize particular locations in visual space is the enhancement of activation in the cortical representation of that part of space. In the dorsal attention system the enhancement only occurs for locations in the opposite side of space.

Locating targets

An object catches our attention either because it is salient or because we are looking at it. In either case, if the object is in peripheral vision we move our eyes so as to bring it into central vision. One way of assessing the ability to detect objects is the letter cancellation task. A display of different letters is presented and the person is asked to cross out or cancel all the exemplars of a particular letter, for example the letter T as in Figure 8. This task is typically used as a diagnostic test for the clinical syndrome of spatial neglect.

This is a strange and intriguing phenomenon, and indeed it was hearing about it in lectures at university that attracted me into neuroscience. The patients act as if things on the left hand side of space are not there. So they might neglect the food on the left hand side of their plate or they might fail to wash the left hand

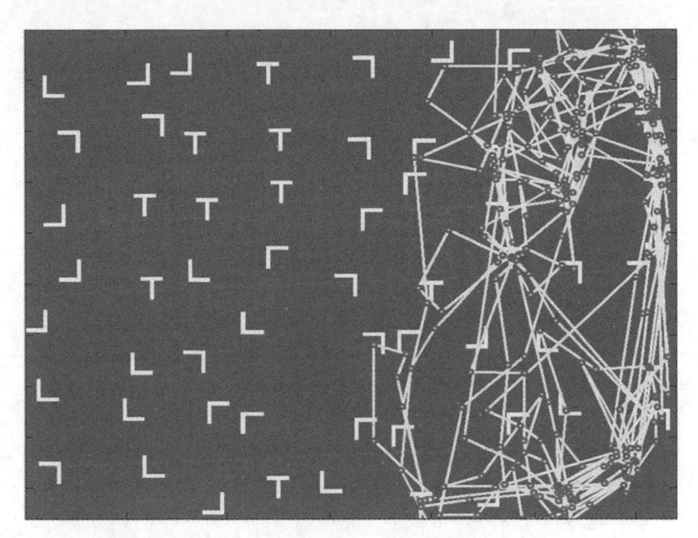

8. Paths of eye movements in a patient with left sided visual neglect performing the letter cancellation task. The target letters are the 'T's.

side of their face. When asked to draw a clock, patients with neglect typically omit the numbers round the left or crowd them round the right hand side. On the letter cancellation task they fail to cancel the letters on the left hand side of the display; and Masud Husain has shown that remarkably their eye movements are almost entirely confined to the right hand side of the display (Figure 8).

Neglect most often results from strokes that involve the right parietal cortex. In a stroke the blood supply to an area is cut off, either because of a blockage or leakage in an artery. The effect is that the neurons in the area that is supplied by that artery die through lack of oxygen and glucose. The damage can be detected by using a CAT (computerized axial tomography) or MRI scan to visualize the anatomy of the brains, and it turns out that the lesions usually include the inferior parietal cortex, that is, the cortex below the intraparietal sulcus (Figure 7). Within this area it has been suggested that the critical region is the cortex at the junction between the parietal and temporal lobes, referred to as the temporo-parietal junction or TPJ (Figure 7).

The inferior parietal cortex and TPJ form part of the 'ventral attention system'. As defined on the basis of resting state covariance the TPJ is interconnected with the anterior insula (Figure 7). As expected given its location, the TPJ is also connected with the temporal lobe, and in particular the tissue in the superior temporal sulcus and the inferior temporal cortex (Figure 7). The implication is that it forms a critical node at which the temporal and parietal cortex can interact.

Target detection tasks such as the letter cancellation task provide one situation where that interaction is required. The ventral visual stream is involved because the letters are identified on the basis of their shape; and the parietal cortex is involved because some of the targets will appear in peripheral vision.

Gordon Shulman and Maurizio Corbetta devised a task with some of the features of the letter cancellation task. Individuals had to detect a particular object amongst an array of objects, and the objects were presented on each side of the central point. On each trial a cue specified which side they were to attend to and the target object always appeared amongst the objects on the attended side, whether on the left or right. The attention was covert because the observers were required to keep their gaze on the central point.

People were scanned while they performed this task and there were two main findings. The first was that, when targets were detected, there was activation in the right inferior parietal cortex and TPJ irrespective of whether the targets were on the left or the right. By contrast the left inferior parietal cortex and TPJ were only activated when the targets were detected on the right.

The second finding concerns what happened when the cues were presented to tell the observers which side to attend to. In the dorsal as well as the ventral attention system the activations were greater when attention was shifted to the other side rather than being directed to the same side as on the previous trial. But uniquely this was true for the right TPJ irrespective of the side to which it was switched; by contrast the signal in the left TPJ was only greater for switches to the right.

Thus, though they are interconnected, the dorsal and ventral attention systems differ both in their organization and function. The cortex in the intraparietal sulcus of each hemisphere is involved in attending to the opposite side of space. By contrast the right inferior parietal cortex and in particular the TPJ are engaged irrespective of whether attention is to the left or right side of space. In other words they cover the whole of visual space.

Now consider what happens when a large stroke affects either the left or right parietal cortex. After damage to the left parietal cortex, the right inferior parietal cortex and TPJ are still available for detecting targets whether they appear on the left or right. But after a large parietal lesion on the right, the left parietal cortex can only support detection of targets on the right. A neat explanation of spatial neglect, though in science neat doesn't necessarily mean correct.

Identifying targets

As already mentioned, it is the temporal lobe that is involved in recognizing the targets. This involves matching each item against a template so as to decide whether it is or is not one of the target items. In an experiment in Sabine Kastner's laboratory, photos were shown of cluttered scenes; in some there were people, in others cars, and in yet others trees. However, the critical displays were ones in which both people and cars appeared; in these the people were the targets (positive) and the cars the distractors (negative). The photos with trees alone were not relevant to the task (neutral).

A multi-voxel pattern analysis was used to identify from the patterns of activation what class of objects was being shown. This is the method that was briefly described in Chapter 2. The analysis was performed on the activations in an area of the temporal lobe that is involved in the recognition of objects.

The finding was that from the pattern of activation it was easier to identify people (targets) than trees (neutral), but more difficult to identify cars (distractors) than trees (neutral). This suggests that while the representation of the targets was enhanced, that of the distractors was suppressed. The effect is that in a visual search task the distinction between targets and distractors is maximized.

But this does not explain how the visual system is set up to detect the targets. So in another imaging experiment by the same group photos of cluttered scenes were again presented; some included cars, some people, and some both cars and people. The difference is that on some trials the task was to detect the people and on others the cars.

A multi-voxel pattern analysis was again carried out. However, this was applied to the activations *after* the instruction had been given but *before* the photos had been presented. It proved possible to identify what the targets were for that trial by analysing the data from an area that was shown to be involved in the recognition of objects. What this means is that if the people were told that the targets were cars, the category of car was set up as a template; when the photos were presented, the objects could be compared with that template so as to decide whether they were or were not cars.

Focused attention

In a visual search task the distractor items are deliberately introduced so as to make it difficult to detect the targets. But there are other situations in everyday life in which we try to focus on one task in the face of distraction. For example, we are trying to remember something while someone else is talking.

To study this situation, Christian Büchel scanned people while they had to remember a series of tones and at the same time they were presented with distracting pictures of objects. The finding was that distraction reduced the activation in the LO complex, and this was especially so when the auditory memory task was difficult. This suggests that information in the irrelevant visual stream was inhibited.

It is this effect that accounts for our ability to distract ourselves when we are in pain or are experiencing tinnitus in our ears. In another experiment by the same team a painful stimulus was applied to the backs of people while they were scanned. As in the experiment just described the people were required to carry out a memory task at the same time. Mental distraction led to a decrease in the activation in the sensory pathway in the spinal cord, and the people also reported that they felt less pain.

The signals that cause this reduction of pain originate in the prefrontal cortex. It sends connections back to the higher sensory areas (Figure 9). Signals that pass through these pathways are referred to as 'top–down' signals so as to contrast them with the 'bottom–up' signals that arrive via the primary sensory areas. In the case of pain these signals appear to evoke a release of opiates in the system and it is this release that moderates the pain.

The reason why these signals originate in the prefrontal cortex is that it is this area that is involved in setting up the task. At the

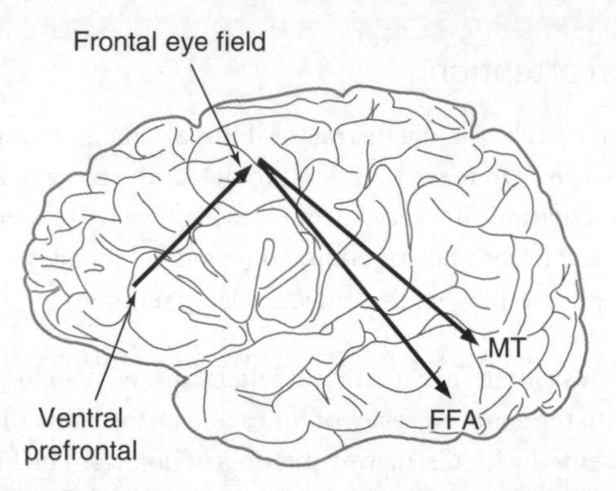

9. Top-down pathways from prefrontal cortex to the MT complex and the fusiform face area (FFA).

beginning of an experiment the people are told what to do, and Marcel Brass has shown that a preparatory signal can be detected in the ventral prefrontal cortex after the instruction has been given and before the task begins.

In a typical experiment the task instruction will specify what material is or is not relevant for the task. So Katsuyuki Sakai and I devised a task in which a series of four letters was presented at the same time as a series of four squares. The task was either to remember the letters or to remember the locations of the squares. At the beginning of each trial a verbal instruction was presented to tell the people which of the two tasks to perform. After the instruction but before presentation of the items there was an activation in the ventral prefrontal cortex, and this continued during the interval until the letters and locations were presented.

This continuing activation reflected the setting up of the task; in other words it constituted 'set' activity. The evidence for this was that on spatial trials the set activity was more closely related to the activation in areas involved in spatial processing; whereas on letter trials it was more closely related to activation in areas involved in verbal processing. This supports the claim that it was the prefrontal cortex that set up the modality that was relevant to the task.

But though suggestive, this evidence is not conclusive. The fact that A is related to B does not prove that A causes B. The way to find out is to manipulate A and see if there is a change in B. In an experiment in John Driver's laboratory transcranial magnetic stimulation (TMS) was used to stimulate the frontal eye field (Figure 9). The effect was to induce a signal that passed through top–down pathways to the higher sensory areas (Figure 9). This signal could be detected because the experiment was performed in the scanner and so changes in activation could be measured in the sensory areas.

During scanning the individuals were presented with faces made up of moving dots: one task was to tell the gender of the face, the other the direction that the dots were moving. If the task was to attend to the face, prefrontal stimulation led to enhanced activation in the fusiform face area (FFA), an area in the inferior temporal cortex that is involved in the discrimination of faces (Figure 9). If the task was to attend to motion, stimulation of the prefrontal cortex led to enhanced activation in the MT complex (Figure 9).

So top–down signals can exert a selective effect in three ways. They can set up the task by enhancing processing in the relevant sensory stream. They can set up the targets by creating a template against which they can be matched. And, if a task is performed in the face of distraction, they can inhibit processing in the irrelevant stream.

Divided attention

In the experiments described so far the people were required to attend selectively, but in everyday life we sometimes engage in multitasking. We use a mobile phone while driving the car. We assume that this is safe if the phone is hands-free because we can keep both hands on the steering wheel. However, we still have to keep an eye on the road at the same time as responding on the phone.

We can simulate the situation in the laboratory by requiring people to respond to visual and auditory cues at the same time. In one experiment the people had to press different fingers depending on the shape presented and also to repeat the words that they heard. The finding was that there was a penalty to pay in that it took longer to respond when doing both tasks together than when doing either task on its own.

But why should this be? Chapter 2 stressed that the brain can perform many operations in parallel. Different areas are involved in analysing what we see or hear (Figure 10), and so we can see at the same time as we hear. And different areas are involved in moving the fingers and speaking (Figure 10), and so we should be able to press a finger at the same time as talking.

The reason why there is a problem in doing a visual/manual and auditory/vocal task at the same time is that there is a stage at which the same area is engaged by both tasks. Paul Dux and René

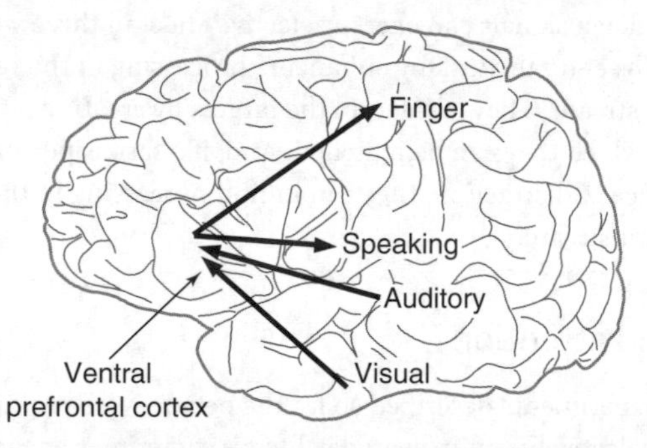

10. Ventral prefrontal cortex as a bottleneck during visual/manual and auditory/vocal tasks.

Marois have shown this is true for the ventral prefrontal cortex (Figure 10). It is here that the sensory input is transformed into the motor output: one component of the activation reflects the sensory input and another the motor output.

So the interference occurs at the point at which a decision is taken as to the appropriate response. It is here that the processing is serial. This would account for the increase in the times taken to react when the two tasks are performed simultaneously.

Of course someone could object that the fact that the same *area* is involved in the transformation does not mean that the same *neurons* are involved. But there is an indication that they probably are. This comes from an experiment in which Paul Dux and René Marois reversed the mappings so that the individuals had to speak in response to visual cues and move different fingers depending on the word they heard. The finding was that the ventral prefrontal cortex was engaged by both tasks as before. This suggests the reason why the different tasks are routed through the same area. If the system is to transform *any* input into *any* output it needs a matrix of *interconnecting* neurons. Given such a matrix, routing through the same area allows for maximum flexibility.

But is all this relevant for driving? After all, driving is an automatic skill, and we know that, if people are trained on a visual/manual task for several days, there is less interference when they do this task at the same time as repeating words that they hear. The reason is that, as Paul Dux and René Marois have also shown, the activation in the ventral prefrontal cortex is over more quickly when a task has been overtrained. This means that though the two tasks are processed serially, there is less cost in terms of time.

The implication is that under normal conditions it is indeed safe to carry on a conversation while driving because you will respond automatically to the road ahead. The danger comes when you are talking on the mobile but then have to decide how react to a *novel* situation. If people are newly trained on a visual/manual task, they sometimes move the wrong finger when they have to repeat words at the same time. Under pressure, there is less time to decide on the appropriate response.

So beware the unexpected. Someone steps out into the road and you have to decide which way to turn to avoid them. The delay means that there could be a fatal accident.

Answers

1. The parietal lobe represents visual space. In the dorsal attention system the left hemisphere represents right space and the right hemisphere left space. But in the ventral attention system the right inferior parietal cortex and TPJ represent both left and right space whereas the corresponding areas on the left only represent right space. So after a large lesion to the right inferior parietal cortex and TPJ there is no remaining representation of left space, and the result is that the patients fail to take notice of things on the left.
2. Of the incoming stimuli, only some are relevant at any one time for what you are doing. There are top-down signals that select between the sensory streams and they do this both by enhancing the attended stream and inhibiting the unattended one. So it is possible to distract yourself, for example from pain, by carrying out a difficult mental task; the effect is to diminish your awareness of the pain.
3. The processing of the different sensory streams occurs in separate areas; so you can, for example, identify pictures and hear words simultaneously. And this is also the case for different responses; so you can, for example, use your hands while speaking. It is at the stage where the sensory input

must be transformed into the appropriate output that interference occurs when multitasking. The reason is that the different tasks are routed through the same area; the result is a bottleneck such that only one decision can be taken at any one time. The reason why it is dangerous to use a mobile phone while driving is that, because of this bottleneck, there will be a delay in deciding how to react if unexpected events occur on the road ahead.

Chapter 4
Remembering

Questions

1. How can we explain amnesia for the past, yet retention of knowledge learned at school?
2. Why do we begin to forget people's names when we are in our fifties?
3. Why do people with Alzheimer's disease have difficulty in finding their way?

Visual attention is needed because objects are located in cluttered scenes. You are looking for your keys in the kitchen; and the cooker, the storage cabinets, and the sink form part of that scene. The problem is that, as you search, you cannot see where your keys are.

Now change the scenario. You are not in the kitchen and the problem is that you cannot *remember* where you put your keys. You search the house, looking from room to room and from one piece of furniture to another. No luck. But as you approach the kitchen you suddenly recall that you put them down on the worktop. The sight of the kitchen has been enough to remind you. So the scene (the kitchen) prompts your memory (where you put the keys).

There is something else about that moment of enlightenment. It is not just that you know where your keys are: it is as if 'it all comes back to you'. These memories that you recollect from your past are referred to as 'episodic' or 'autobiographical' because they are memories of events in your own life.

Scenes and places

Given that scenes can prompt memories, the first thing to find out is how scenes are constructed. Unlike small objects, which we can fixate in central vision, scenes are extended and so they fill our visual field. There are distinct pathways for central and peripheral vision and the latter terminate in a structure called the parahippocampal cortex that lies towards the inner surface of the temporal lobe (Figure 11). As expected, the parahippocampal cortex is activated if people are scanned while they view pictures of scenes or large objects such as houses that could serve as landmarks.

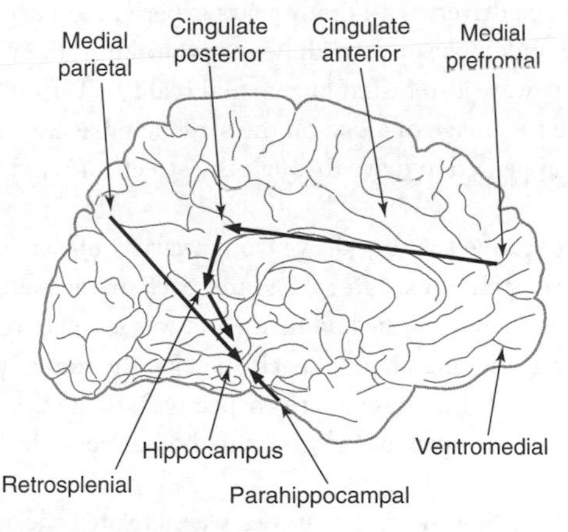

11. Connections of the hippocampus.

Resting state covariance can be used to study the connections of this area. The finding is that the parahippocampal cortex is connected with the inferior parietal cortex and with two areas on the medial or inner surface of the brain, the retrosplenial cortex and the hippocampus (Figure 11). These areas also receive a direct input from the medial and inferior parietal cortex (Figure 11), and it is known that there are several parietal areas in which maps of external space can be identified. So the hippocampus receives information about the external scene and surroundings.

The hippocampus represents where we are in those surroundings. Demis Hassabis and Eleanor Maguire scanned people while they learned their way around a virtual reality environment on the computer screen. A multi-voxel pattern analysis was then used to see if it is was possible to tell where the person was in the environment; and it turned out that indeed this could be done from the pattern of activation in the posterior part of the hippocampus.

You need to know where you are if you are to find your way. In a supremely elegant experiment Hugo Spiers and Eleanor Maguire recruited taxi drivers with many years experience of driving through London because, of all people, they know their way. The taxi drivers were also tested in a virtual reality set-up. They controlled the image of a taxi on the screen and as they moved the taxi, so it appeared to drive through the streets of London.

They were scanned as they drove from a variety of locations to particular destinations. After the scans, each driver was shown the video of the routes that he had taken and was asked to remember what he was thinking about at each stage on the route. It was then possible to relate the thoughts retrospectively to the dynamic brain images that were taken at the time that he was actually driving.

The *only* time that the hippocampus was activated was when the drivers were shown a familiar starting location and told the

12. St Paul's Cathedral and the Bank of England.

destination to which they had to drive. So this was when the drivers knew that they were at location A and had to plan the quickest way to reach location B. Given that the hippocampus codes for where we are, the experiment tells us that this knowledge serves as a *context* or reminder for a spatial direction or route.

If so, someone with damage to the hippocampus should have trouble in finding their way around. So the same authors contacted a taxi driver who had suffered from an infection of the brain that damaged his hippocampus in both hemispheres. They then assessed the effect by testing him in the same virtual reality environment.

After his stroke he was completely lost. Take as an example his attempt to drive from St Paul's Cathedral to the Bank of England (Figure 12). It is only a short drive, yet he veered off and never regained his sense of direction. His route took him all round the place. Yet he had had over forty years' experience of driving through London.

From places to recollection

There is another effect of damage to the hippocampus and this is even more disabling. This is that the patients are severely amnesic for past events in their life. Faraneh Vargha-Khadem has described a patient called Jon who suffered a small stroke in the posterior cerebral artery at birth and this resulted in the loss of neurons in the posterior hippocampus. The lesion is incomplete, but he is still very poor at remembering events such as who has visited recently or where he has been for a holiday. And when asked, he says that he is unable to conjure up the experience or relive the events.

There is independent evidence that the hippocampus is involved in remembering events. This comes from studies that use brain imaging. For example, Donna Addis, Daniel Schacter, and colleagues scanned people and showed that the hippocampus was activated when the people recalled events from their life.

So the hippocampus is activated in two situations, when people think about how to get from A to B and when they recall past events. But it is not obvious how a system for finding your way should also be able to support recall of episodic memories. Indeed, when John O'Keefe first showed that there are neurons in rats that code for the location of the animal, many were sceptical—including the author. John O'Keefe went on to win the Nobel Prize. So much for scepticism.

At the time the problem was that it was not clear what coding for place had to do with remembering events. Place is a simple matter of location. But episodic memories are not vague memories of *where* something happened; they remind us of *what* happened and *when* it did so. There is still no agreement about the relation between coding for place and memories of this sort, and so what follows is a half-educated guess.

The suggestion is that the key to the puzzle is to ask *why* the animal wants to find a particular place. In their natural surroundings the reason is that that is where food, a mate, or their offspring are. They know the association (*what* and *where*) because they have found food there before, mated, or given birth. That is why they return.

But this knowledge is not sufficient for survival. The world changes: water is available in different places at different times, predators move about, and when in danger females move their young. So it pays animals to know not only *where* they last found water, spotted a predator, or deposited their young but also *when*.

All this is theoretical. No proof has been given that animals can remember these things or that they can do so on the basis of a single observation or experience. To answer this objection Nicky Clayton has studied scrub jays because they cache food and so it is possible to test their memory for where they have done so. The experiment was ingenious. The birds were given meal worms and peanuts to bury. If tested soon after they retrieved the meal worms, but if tested five days later they retrieved the peanuts. The assumption is that they did so because they knew the meal worms to be perishable and the reason why they avoided them after five days was that they remembered when they had cached them. The conclusion is that they remembered *what* foods they had cached (meal worms or peanuts), *where* they had cached them, and *when* they had done so.

Impressive, but not a patch on chimpanzees. Charles Menzel carried out an experiment with two different chimpanzees. They watched while someone hid ten bags of almonds in a large wooded field. The bags differed in the number of almonds and in the effort required to obtain the food: some of the almonds were in shells and some were not. After twenty to twenty-five minutes the chimpanzees could then guide another person to where each bag was hidden and also work out the quickest way to find them, calculating the optimal path that took into account the relative value of the items. The animals were tested over many days, and so they had to remember where the almonds had been hidden on any particular day.

A neuroscientist might object that the experiment has not shown that the hippocampus is necessary for recall. And philosophers *have* objected that there is no evidence that the animals can recollect in the sense of 're-experiencing' the memory. Fortunately for philosophers we can't ask the animals.

But there is a way of finding out what happens in the human brain during recall, to see whether there is a similarity between the state of the brain during the original event and at recall. So a group in the laboratory of Neil Burgess scanned people while they viewed videos and later while they recalled them. The advantage of videos is that the experimenter can control the content of the events. It also means that it is easy to test whether the people are able to recall details of these events accurately.

The videos were shown several times during scanning, and the more often the same film was repeated, the greater the detail the people could report. A pattern analysis was used to compare the activations when viewing the videos and when recalling them a week later. To test for specificity, the analysis checked to what extent the state of brain was more similar when viewing film 1 and recalling film 1 than when viewing film 1 and recalling film 2.

There were many areas in which the state at viewing was reinstated at recall. These included the hippocampus, retrosplenial cortex, posterior cingulate cortex, and the medial and inferior parietal cortex (Figure 11). And the degree of similarity in the posterior cingulate cortex was related to the amount of detail that was recalled. It is reinstatement that underlies the subjective feeling of re-experiencing the event.

The limitation of the experiment is that it does not show that the hippocampus is responsible for the reinstatement in the cortex. To demonstrate this, it is necessary to interfere with the workings of the hippocampus. So in one experiment mice were given a mild electric shock. When the animals were later put back in the same situation, many of the same neurons were reactivated in the neocortex. But this no longer occurred if the activity of cells in the hippocampus was temporarily suppressed.

The implication is that the mechanism of reinstatement is common to mammalian memory. So what, if anything, is different about human memory? There are three candidates and of these the first is clear-cut. We live in a technological world and this means that there is a much greater richness of things that we do. Our repertoire is not restricted to simply carrying food about.

There is a second potential difference. This is that, when trying to think where we left our keys, we are aware of letting our thoughts wander until simply imagining the kitchen unlocks the memory of what we did there. People have been scanned while their minds are free to drift, and there is extensive activation on the medial surface: this includes not only the posterior cingulate cortex and retrosplenial cortex but also the medial prefrontal cortex (Figure 11). The same system is also active when chimpanzees are at rest, but there is no way of knowing what, if anything, is on their minds.

The third candidate relates to the fact that we can retrieve memories from the distant past and imagine the distant future. And the hippocampus and other medial areas are activated when people review events from the past or imagine ones in the future. Bertrand Russell once commented that a chimpanzee cannot think about what it is going to have for breakfast tomorrow. Witty but unproven. In our case it helps, of course, that the calendar provides us with a scale in days, months, years, and decades so that we can place events along this scale.

It allowed Heidi Bonnici and Eleanor Maguire to ask people for their memories from two weeks ago or ten years ago. During scanning, a multi-voxel pattern analysis was used to see if it was possible to distinguish between different memories. This proved possible when analysing the activations in the hippocampus, retrosplenial cortex, and ventromedial prefrontal cortex (Figure 11, p. 43). In the last area, in particular, the degree of accuracy was better for the memories that were remote. One could speculate about possible differences between memories that are reinstated from different periods of time. But fortunately an introduction is not the place to do it.

So in summary, the suggestion is that the hippocampus provides the spatial context for events, whether the context is actually present or only there in the imagination. The events are autobiographical or personal because we were present at the time. And recall depends on the process of reinstatement in the cortex. It is this that we experience as reliving the event. Indeed, if the events were traumatic we describe them as 'flashbacks' and it may be difficult to get rid of them.

Semantic knowledge

If the specific role of the hippocampus is to enable us to retrieve memories, intuition tells us that damage to the hippocampus should also interfere with learning at school. After all, factual learning also requires memory and teachers often test their pupils to see if they remember what they have been taught. Yet Faraneh Vargha-Khadem found that surprisingly Jon had been able to learn at school.

So there is a fundamental difference between remembering events in your life and remembering facts that you have been taught. This distinction is marked in the literature by contrasting 'episodic' with 'semantic' memory. It is one thing to remember where you put your keys, another to know what keys can be used for.

So Rik Vandenberghe and Cathy Price scanned people while they were tested for their knowledge of the properties of things such as tools. The experiment has already been mentioned in Chapter 1. In the problem illustrated there in Figure 3 (p. 6), individuals had to

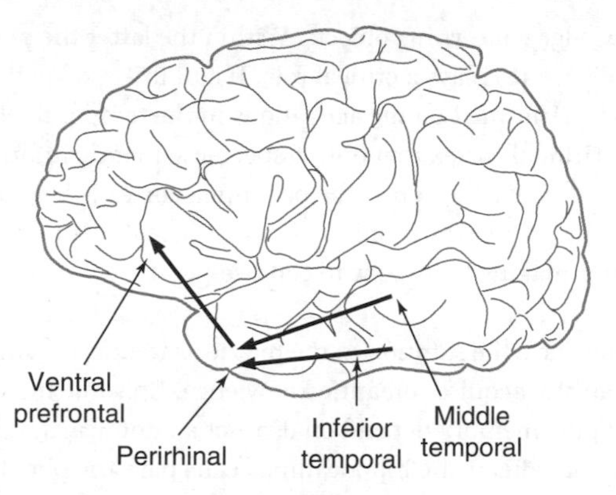

13. Semantic system as shown by fMRI.

judge whether a spanner or a saw is related to a pair of pliers. The answer is the spanner because it can be used to grip. The problems were either given in the form of pictures, as in Figure 3, or in words. In the latter case the word 'pliers' was shown above and the words 'spanner' and 'saw' below.

Irrespective of whether the problems were presented in words or pictures, there was activation in the middle and inferior temporal cortex extending to the perirhinal cortex, which lies at the temporal pole (Figure 13). There was also activation in the ventral prefrontal cortex with which these are connected. The activations in common for the pictorial and verbal versions were all in the left hemisphere.

So why were they in the left hemisphere? The most obvious reason is that it is the left hemisphere that is specialized for language (Chapter 8), and most of what we know comes from being told it or from reading it in books. That is why it is called 'semantic knowledge'. Even if we have never actually used a spanner we still know what to do with one.

There is, then, a clear distinction between a medial system for remembering events in our life, and a ventral and lateral system for knowledge concerning objects. Within the latter the perirhinal cortex (Figure 13) plays a critical role. If a multi-voxel pattern analysis is performed on the activations in the temporal lobe, it is in the perirhinal cortex that object-specific semantic information is represented. Furthermore, the perirhinal cortex is increasingly engaged the greater the number of semantic features that objects share, thus making them easy to confuse.

So we now have the solution to the puzzle as to how patients such as Jon can still acquire semantic knowledge. Episodic memory and semantic memory depend on distinct anatomical systems. A stroke that affects the hippocampus can spare the perirhinal cortex because different arteries supply the two areas.

Ageing

As we age, our memory is not what it was. In healthy individuals the signs are annoying but in dementia they are disastrous. Sadly the suggestion is that if we lived long enough we would all progress to dementia in the end.

One early complaint that people make is that they are having trouble in remembering the names of people. This can hit us in our fifties. The problem cannot simply be that the names of people are arbitrary because so is the name of a cow; each language has a different word for this.

Our knowledge of the meanings of words forms part of our semantic knowledge and this degenerates in people who suffer from 'semantic dementia' in old age. The syndrome results from the loss of the temporal and prefrontal tissue shown in Figure 13; and critically this includes the perirhinal cortex. Damage to the perirhinal cortex has been shown to lead to problems in naming things, but this problem is worse the more they are perceptually and semantically confusable.

Now consider the names of people. Different people share many visual features: they have eyes, noses, and so on. But there are also many other properties that the people that we know have in common: they may be colleagues or they may be in our list of email contacts. So all these people are confusable and yet the name that we are trying to remember is only attached to one of them. The result is that remembering personal names is one of the most taxing semantic tasks and thus one that is particularly sensitive to ageing.

But as we age we also complain of forgetting events, such as where we put our keys. In the case of Alzheimer's disease it is possible to detect a loss of volume in the hippocampus seven years or so before it becomes obvious that there is a severe memory problem. Of these the most obvious is that the people are unable to retrieve memories of their life. But it is also common for them to be confused about where they are or how they got there and the reason is that, as described earlier, the hippocampus represents the location of the individual.

One study assessed the ability to follow routes by showing a video of a car driving along the road and taking left or right turns at particular landmarks. The person was then shown the film for a second time and at each intersection asked which way to turn. Patients with Alzheimer's disease made many errors whereas patients with semantic dementia had little trouble.

But there was a second test and this was for what the road signs meant. Here the positions were reversed: it was now the patients with semantic dementia who were in difficulties. This is another demonstration that remembering is not unitary. There are dissociable systems for memories of what we have personally experienced and memories of what we have learned or been told.

It is, however, a sad fact that, as Alzheimer's disease progresses, the pathology becomes widespread until it spreads to the perirhinal cortex and elsewhere. The result is that both episodic memory and semantic knowledge become severely compromised. It may be lucky for scientists that there are patients with natural lesions that allow them to distinguish between systems. But for patients with Alzheimer's disease nature shows no mercy.

Answers

1. The hippocampus is critical for the retrieval of episodic memories. These are personal memories of events from your life. The hippocampus represents the spatial context in which the event occurred and thus acts to prompt recall of that event. This system is anatomically separate from the semantic system and so it is possible for a stroke to affect one but not the other.
2. As we age we tend to forget names. This is not because they are arbitrary but because different people share many visual and semantic features but the name is attached to only one person. The perirhinal cortex is especially taxed when things are confusable in this way, and it is the perirhinal cortex that is most compromised in semantic dementia.
3. The fundamental role of the hippocampus is to represent the location of the person in space. It is engaged whether the person considers what route to take or conjures up an event from the past. In Alzheimer's disease there is a loss of tissue in the hippocampus; and the result is that, as well as having trouble retrieving episodic memories, the patients can fail to know where they are or where they have been.

Chapter 5
Reasoning

Questions

1. Why do different intellectual abilities tend to cluster together?
2. Do we think in language?
3. How come human beings are so intelligent?

We distinguish between knowledge and ability. Someone can be well taught but unintelligent and another clever but uninformed. It is for this reason that, when I was a student at Oxford University, it was possible to get a mark of alpha/gamma ($\alpha\gamma$) in an exam. The implication of the alpha mark was that the answers showed that the candidate was full of ideas and the implication of the gamma that the person knew very little.

One way of assessing the ability to think is to give an intelligence test. The best of these present problems that cannot be solved simply on the basis of past knowledge. This can be done by avoiding the use of words and showing designs instead. The Raven's Matrices is a typical example of such a non-verbal IQ test; Figure 14 shows a typical problem.

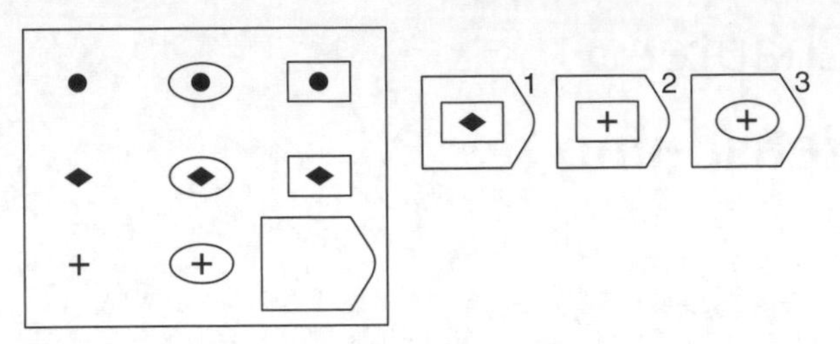

14. Problem taken from the Raven's Matrices. The person has to choose between designs 1, 2, and 3 on the right.

To solve the problem you look at the first line and try to detect the rule or sequence. You then inspect the second line to see if the same rule applies. If it does, you are then in a position to complete the third line by selecting the item that is appropriate. Tests like this are referred to as tests of 'fluid intelligence' because they test the ability to think anew rather simply retrieving past knowledge.

Non-verbal problem solving

People have been scanned while they solved problems taken from the Raven's Matrices. For the more demanding problems the activations were in the parietal and dorsal prefrontal cortex (Figure 15, upper line). These areas are known to be interconnected.

An alternative way of testing fluid intelligence is to present sequences of letters: here the task is to detect the rule governing the sequence and then to complete the sequence. John Duncan specifically compared problems of this sort with problems involving designs; and the same parietal and dorsal prefrontal system was activated irrespective of the type of material or rule.

These results are important because non-verbal tests of this sort provide the most reliable assessment of general intelligence,

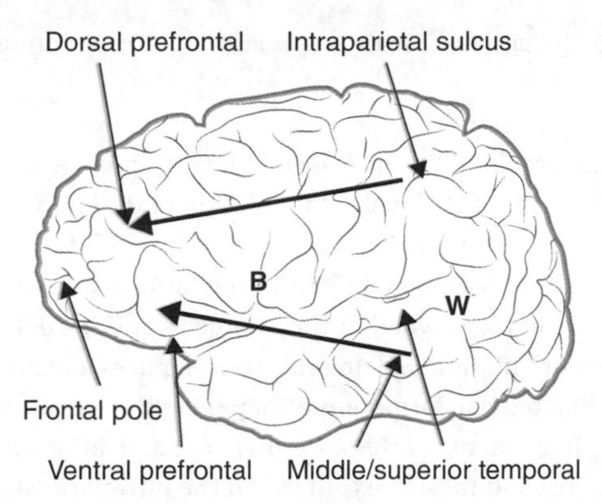

15. Areas involved in non-verbal and verbal reasoning. B = Broca's area, W = Wernicke's area.

sometimes referred to as *g*. The problem with verbal tests is that they can be biased in favour of particular cultures or those with a particular education. That is why tests with shapes or symbols are thought to be more 'culture fair'.

One way of testing whether they really are is to study the effects of lesions. If these problems can be solved without relying on past knowledge, lesions in the semantic system should have no effect. Chapter 4 has shown that the temporal lobe forms part of the semantic system; and, as predicted, patients with lesions in the temporal lobe are unimpaired on tests that use shapes or symbols. The patients who are impaired are those with parietal or prefrontal lesions, and the impairment is more severe the larger the lesion.

Spatial relations

There is a reason why it is the parietal and dorsal prefrontal system that is critical for performance on tests of fluid intelligence. This is that they typically involve rules that involve the manipulation of spatial relations. And, as described in

Chapter 3, the parietal cortex is specialized for the representation of space.

Some of the problems on the Raven's Matrices require the ability to appreciate rules that involve the rotation of designs or elements of the designs. There is a direct way to test this ability: this is to present shapes or letters in different orientations, and to ask the observers to judge whether they would be identical if shown in the same orientation. Solving this task requires a mental manipulation or transformation. When people are scanned while they make judgements of this sort, there are activations in the parietal cortex and these are centred on the intraparietal sulcus (Figure 15).

Other tests of fluid intelligence assess the ability to relate the position of items in a series, as, for example, in the case of letter series. So Guy Orban presented pairs of letters and the observers had to say which comes later in the alphabet. In a second condition the observers had to judge which of two numbers was larger. In both cases there were increases in activation in the cortex in the intraparietal sulcus.

The reason for the activation in common is that there is a relation between space and distance, and judgements for letters or numbers are made on the basis of the distance apart in the series. The letters and numbers are viewed as if on a line, and in the case of numbers this is referred to as the 'number line'.

Stanislav Dehaene and his colleagues have shown that there is a relation between size, distance, and number. They presented the words for two numbers and the observers were asked which was the larger. The nearer the two numbers the greater the activation in the cortex in the intraparietal sulcus. And the same result was found when the observers had to make judgements of size. So the reason why the parietal cortex is critical for fluid reasoning is that it codes for a common set of metrics and these all relate to space.

Thought and language

Though tests of fluid reasoning are not presented in language, it is a common belief that when we reason we do so in language. Imaging provides a way of finding out if this is true. We are all aware that we can sometimes 'hear ourselves think' and this ability is referred to as 'inner speech'. Children sometimes chatter to themselves as they tackle problems but as responsible adults we have learned to suppress overt speech so as to avoid embarrassment.

We can test whether we are really 'speaking' because we know where there are activations when we speak or hear speech (Chapter 8). The areas involved are Broca's area (Figure 15, B) and Wernicke's area (Figure 15, W). These are named after the neurologists who first showed that they are critical for language.

These two areas form part of what is called the 'phonological' as opposed to the semantic system. When people silently repeat a word to themselves in the scanner there is activation in Broca's area and in the neighbourhood of Wernicke's area. So it is as if we are indeed hearing ourselves speak.

Now we are in a position to test whether we are talking to ourselves when we think through logical problems. Vinod Goel and Ray Dolan devised tests of deductive reasoning. All involved syllogisms such as 'all apples are red fruit; all red fruit are nutritious; so all apples are nutritious'. The three statements were presented on the screen, one by one. The task was to judge whether the conclusion followed from the premises. During the scans there were no activations in either Broca's or Wernicke's area; in other words the people were not using inner speech to solve the problems.

Instead the activations were in the middle temporal cortex and the ventral prefrontal cortex in the left hemisphere (Figure 15, lower line), and as described in Chapter 4 these form part of the semantic system. The reason why this was engaged was not that the people had to understand the words, because the experimental design included a control condition in which the last sentence was irrelevant for the argument. In the example given above, it could be 'no napkins are white'. In this condition it is also necessary to understand the words, but anyone can see that the syllogism is invalid; we might call it a 'pseudo-syllogism'.

One reason why the semantic system was activated for the genuine syllogisms was probably that the people were appealing to their knowledge of objects. But this is not the way to decide whether the syllogisms were valid. The syllogism above would be valid even if you knew that some berries are poisonous.

The syllogisms hold because of logic. Thus, the argument can also be presented in symbolic form as in: 'all As are Bs; all Bs are Cs; so all As are Cs'. So Vinod Goel included syllogisms in this form. Irrespective of whether the syllogisms were presented in words or symbols, there were activations in the parietal cortex, though the activations were greater for the syllogisms when presented in symbols.

The reason why the parietal cortex was engaged in both cases is presumably that it can represent relationships. These include relations such as equality or difference, larger than or smaller than. And logic depends on relations such as these.

Of course, the fact that the parietal cortex is activated does not prove that the system is critical for successful performance. So Jordan Grafman and Vinod Goel gave logical problems to patients with parietal lesions. The problems involved transitive inference as in: 'Mary is taller than Fiona; Mary is taller than Carol; Fiona is taller than Rosemary; so Mary is taller than Rosemary'. The judgements of the patients with parietal lesions were impaired, even though the inferences were presented in words.

These results explain how it is possible for someone to reason even if they lack the ability to engage in inner speech. A patient has been described with a temporal lobe lesion that left him with no awareness of inner speech. Yet there was no evidence that he was unable to think in everyday life. Though he was not tested on syllogisms, he was tested on the Raven's Matrices and on this he was assessed as having a high IQ. This was possible because his lesion was in the temporal lobe and the parietal cortex was not damaged.

General intelligence or *g*

When assessing intelligence in the clinic, both verbal and non-verbal tests are usually included. The Wechsler Adult Intelligence Scale (WAIS) is a standard test, and in its most recent version it consists of ten core subtests: four are verbal, four visuospatial, and two symbolic. Though the subtests are designed to assess different aspects of intelligence, it has long been known that the results on the different subtests tend to give similar results. It is this finding that has led to the suggestion that people differ in their general intelligence or g.

One of the verbal subtests on the WAIS is called 'Similarities'. In this the person has to decide whether two objects or two concepts are or are not similar. This taps the ability of people to understand relations between categories. The same ability can be tested by presenting analogies and asking if they hold.

For example, in a study in Sylvia Bunge's laboratory the analogies were shown in pictorial form. One depicted the analogy 'painter is to brush as writer is to pen'. In the experiment, there were two ways of assessing whether the people could appreciate the analogies. One was to show them the analogy in an incomplete form and ask the participants to pick the missing item; in this case it would be 'pen'. The other was to show them the complete analogy and ask them whether it was valid.

The associations between a painter and brush or a writer and pen are both semantic ones, like the associations between a spanner and a pair of pliers (Chapter 4). So it is not surprising that there were activations in the left middle temporal gyrus and the ventral prefrontal cortex. However, just as in the case of verbal syllogisms, there was also activation in the parietal cortex. Judging whether two relations are the same is like comparing two quantities or lengths; in other words similarity is judged in terms of equality.

It is because the parietal cortex is involved, whether the tests involve words or not, that the results for the different subtests tend to be similar. As already mentioned, the parietal cortex is interconnected with the dorsal prefrontal cortex (Figure 15, upper line) and this parietal/prefrontal system is known to be engaged across a very wide variety of cognitive tasks, even though they differ in the type of demands that they make. John Duncan has shown this for tasks that tap the ability to hold material briefly in mind, to add up numbers, or to cope with distracting cues when deciding how to respond. Because this system is engaged on such a wide range of tasks he has suggested the term 'the multiple-demand system'; others prefer to call it 'the executive control system'. Whichever, it is on this system that general intelligence depends.

Human intelligence

Though we do not think in language, this is not to say that language is irrelevant for understanding how it is that human beings are so intelligent. Children are taught in language, whether at home or at school. They are told about the properties of objects and how to categorize them. It is in this way that they learn the abstract concepts that allow them to understand the relations between things.

So it is not just our genes that we inherit from previous generations; we also inherit knowledge and understanding. The result is that these are cumulative. You do not have to be Isaac Newton to understand differential equations.

Of course, we could not have this cultural inheritance without specializations in our brain. A crude way of examining these is to assess the degree to which particular areas are expanded. One way of doing this is to fit a 3D MRI image of the chimpanzee brain to a similar image of the human brain. The degree to which different areas have to be expanded to make this fit provides a measure of the degree to which they are proportionately larger in the human brain.

As expected there has been an expansion in Broca's area and in the non-primary auditory cortex in the superior temporal gyrus (Figure 15). These are the areas that are involved in producing and understanding spoken language.

However, not all areas have been equally expanded. The primary sensory and motor areas are no bigger than one would expect given the size of our body. So the fact that the human brain is so large is the result of differential expansion in particular areas. These include the inferior temporal and parietal cortex; but the most dramatic change has been in the prefrontal cortex and the frontal pole in particular. When compared with the rest of the brain, the frontal pole is proportionately twice as large in the human compared with the chimpanzee brain.

The expansion at the frontal pole is particularly intriguing because the frontal polar cortex (Figure 15) is especially activated when difficult problems are given from the Raven's Matrices. It is also activated when people are tested on analogies and asked to judge whether one relation is the same as another. This taps the ability to handle the abstract notion of similarity.

The effect of the differential expansion of particular areas can be judged by relating the size of these areas to the primary areas.

Information reaches the cortex via the primary sensory areas and it is then analysed and categorized through a series of stages. So the processing power of the system is greater the more these areas are expanded in relation to the inputs.

Jeroen Smaers and I have suggested the term 'remapping factor' for this ratio. The human remapping factor for the middle and inferior temporal cortex is 3.5 times that of the chimpanzee and the factor for the parietal and non-primary visual cortex is 2.5 times that of the chimpanzee. So compared with the chimpanzee brain there is much more tissue in the human brain for processing semantic knowledge about objects and for analysing spatial and other relationships.

Since it is the prefrontal cortex that influences action, it is more appropriate to relate it to the size of the motor areas rather than to the sensory input. The remapping factor is an extraordinary 4 times that of the chimpanzee brain. It is this that accounts for the human ability to plan ahead, which we will consider in Chapter 6.

Answers

1. The ability to pass IQ tests of fluid reasoning depends on the dorsal executive system. This is engaged on a wide range of tasks that pose different demands, and this is true whether the material is verbal or non-verbal. It is for this reason that the results of the different subtests on an IQ battery tend to cluster together, so providing a measure of general intelligence or *g*.
2. Language is for communication, not thought. We do not think in inner speech and thinking is not prevented if someone lacks inner speech. Though reasoning with verbal material involves the semantic system, it also engages the parietal cortex. The reason is that comparing relations, as

in analogical reasoning, is similar to comparing lengths or distances.

3. One reason why human beings are so intelligent is that they are taught in a language. This means that they inherit knowledge and understanding, and this cultural inheritance is cumulative. This has been made possible by the expansion of non-primary areas in the human brain. This expansion involves the language areas, the semantic system, and the dorsal executive system. The most dramatic expansion is in the prefrontal cortex, and it is on this that the human ability to plan ahead depends.

Chapter 6
Deciding

Questions

1. Why do we sometimes make absent-minded mistakes?
2. Why do we choose immediate rewards rather than longer-term rewards that are larger?
3. What is the basis for moral reasoning?

If we do poorly on an IQ test it is only our pride that suffers. But in everyday life the consequence of making poor decisions can be a fall. Fortunately we are equipped with the ability to ponder the alternatives before we leap.

However it is no good pondering if we never commit ourselves to a decision. We may be able to think about breakfast tomorrow but in the end we will have to go for either the cornflakes or the puffed wheat. When we do so it will be our prefrontal cortex that does so.

The reader would be right to object that it is people who decide what to eat and not parts of their brain that do so. The statement is clearly metaphorical. But there is a problem with talking in this way, and this is that it suggests that there is a little person or 'homunculus' holed up in the prefrontal cortex. And, of course, no one supposes this to be true. Gilbert Ryle maintained that there is no room for ghosts in philosophy, and there is certainly no room in science.

Banishing the homunculus

The way to banish the ghost is to provide a full account of the information that is processed by the prefrontal cortex and the nature of the influence that it exerts on other areas. In this way it should be possible to document the nature of the transformation from input to output.

The prefrontal cortex stands in a unique position at the top of the hierarchy of processing. We know its connections because of studies of resting state covariance as well as from studies on animals (Figure 16).

The ventral prefrontal cortex receives inputs from all sense modalities. These come from the inferior temporal cortex for

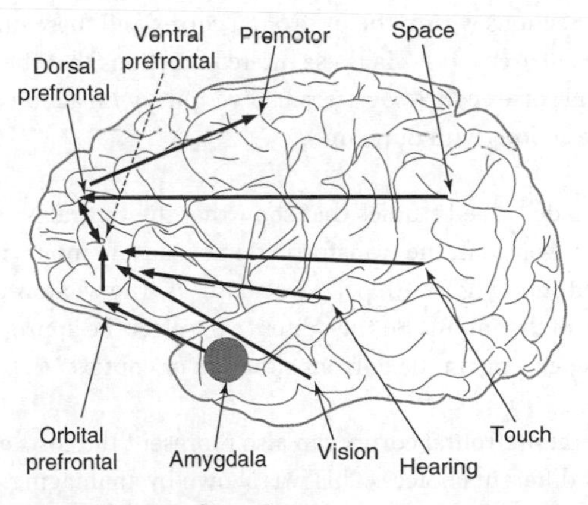

16. Connections of the prefrontal cortex.

vision; from the superior temporal cortex for hearing; from the secondary somatic sensory area in the parietal lobe for touch; and from the inferior parietal lobe for visual space. So the prefrontal cortex is in a position to form a multimodal representation of the current situation in the world outside.

The orbital prefrontal cortex (Figure 16) lies above the orbit and it is connected with the amygdala (Figure 16); in turn the amygdala is interconnected with the hypothalamus (Figure 4, p. 8). The latter is responsive to changes in hunger, thirst, and temperature. So the prefrontal cortex is informed of current needs.

Finally, the dorsal prefrontal cortex can influence action via direct connections with the 'premotor areas'. These lie in front of the motor cortex, hence the term; and it is these that connect directly with the motor cortex. So the prefrontal cortex is in a position to influence actions.

Basing actions on the situation and aims

The connections to the prefrontal cortex from other areas do not all terminate in the same place. But there are strong interconnections within the prefrontal cortex and these link the different subareas. It is via these interconnections that the prefrontal cortex can associate sensory inputs with actions and associate actions with outcomes.

Chapter 3 described studies that show that the ventral prefrontal cortex is involved in the transformation of sensory inputs into motor outputs (Figure 10, p. 38). And this is true whatever the modality of the inputs. So the system can select the appropriate action depending on the current situation or context.

The ventral prefrontal cortex can also represent the costs and values of different choices. This was shown by an imaging experiment by Eire Boorman and Matthew Rushworth. Three pictures were shown on each trial: a face, a body, and a house. During the scans, the probability that picking any particular picture would lead to a 'reward' could be estimated from the previous run of trials. The term reward is typically used in psychology to refer to the food pellets with which animals are trained. But because, as here, people come to experiments well fed, the rewards are more typically points that cumulate on the screen, being converted to money at the end of the experiment.

After the people had made their choice, they were then presented with the two pictures they had not chosen, and asked which alternative they preferred. The finding was that there was a signal in the ventral part of the frontal pole that related to the value of the better of these alternative options. In other words the system can specify the relative values of potential choices.

In this experiment the costs and values were measured in gaining or losing points. But in the real world we often have to compare different types of reward: in the shopping centre the choice might be between buying groceries or clothes. So a metric is needed for estimating the relative value of these items in terms of our current wants. This is done by reference to an abstract scale of value.

One experiment compared the activations for two very different types of reward: taste in the mouth and warmth on the hand. There was a signal that was common to both in the ventromedial prefrontal cortex. The amplitude of this signal was related not to the intensity of the stimulation but to the subjective pleasantness as rated by the people themselves: the more pleasant the sensation, the greater the activation. So the system can compare the value of different rewards.

In summary, the prefrontal cortex is in a unique position in the processing hierarchy. It has access to information about the current situation; it can generate the actions that are appropriate for that situation; it can compare alternative outcomes by reference to a common scale; and it can relate actions to current aims or needs.

Attentive versus habitual action

The advantage of a system with access to information about the current situation and our current aims is that it allows us to alter what we do as either of these vary. To put it another way, it makes possible the moment-to-moment flexibility that is characteristic of human and indeed primate behaviour in general. We move in a complex physical and social world, and by paying attention to each situation we can decide what is or is not appropriate at any one time. Our behaviour can be said to be under 'attentive control'.

But we cannot devote the same degree of attention to all the things that we are doing. As mentioned in Chapter 3, there is a penalty for multitasking. However, it is reduced if one or more of the tasks is well practised.

Consider driving again. Your driving skills are automatic and your route to work is habitual, being the same from day to day. You no longer have to think at every corner which way to turn. The result is that as you drive to work you can be thinking about what meetings you have there for that day or whether divorce is an option.

It would be possible to mimic this situation in the scanner by using virtual reality (Chapter 4). However, science often advances by simplifying the problem. Because the repertoire of movements that can be made in the scanner is restricted, many imaging studies on decision-making limit these decisions to which of several keys to press. Though this makes the experiments appear arid, the logic is preserved. A sequence of turns on a route (left, left, right, and so on) is in principle no different from a sequence of key-presses (keys 2, 4, 1, and so on).

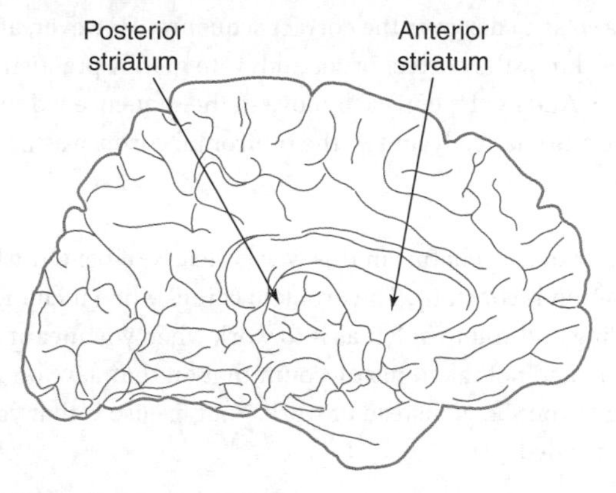

17. Anterior and posterior striatum.

The set-up in an early study by Ivan Toni was that the people had to learn a sequence of eight actions and to do so by trial and error. The key-presses were spaced three seconds apart, and a tick appeared on the screen when that key was correct for that point in the sequence. At first there was widespread activation in the dorsal prefrontal cortex, but by the end of the forty-five-minute period of training this had almost disappeared.

In a later experiment from a different laboratory a sequence was taught in the same way but for two sessions a day and for four days. The scans showed that, by the end, the sequence depended on a restricted circuit that involved the posterior part of the striatum (Figure 17), the medial parietal cortex (Figure 11, p. 43), the supplementary motor cortex (SMA) (Figure 20, p. 85), and the motor cortex (Figure 22, p. 98). These areas are closely interconnected and they provide a mechanism for one movement in the sequence to cue the next.

The reason why the prefrontal cortex was engaged during the initial learning is that the people were learning by trial and error. This means that they had to try different orders in which to press the keys so as to discover the correct sequence. However, after a time they knew the correct order and were merely practising the sequence. And by the end each move in the sequence led to the next one automatically and so the prefrontal cortex was no longer engaged.

The ability to learn habits in this way is clearly of benefit where the situation is constant. The problem comes when it changes. You are driving and you turn left as if to work when you meant to turn right. You had not taken into account that on that day you were heading to the shops instead of work. Your excuse is that you were 'absent minded'.

Donald Broadbent coined the term 'cognitive failures' for these lapses. He told of how once he went to his bedroom to look for something during the day and found himself in bed in his pyjamas. He had not taken into account that it was daytime.

So the lapses occur because habitual actions are run off when they are no longer appropriate. The situation has changed and this means that prefrontal mechanisms should have been re-engaged. To study this effect, people have been scanned while they performed a simple sequence of key-presses (1, 2, 3, 4, 1, 2, 3, 4, and so on). As expected there was no activation in the prefrontal cortex. But all that was needed for the dorsal prefrontal cortex to be re-engaged was for the people to be asked to 'think about' what they were doing.

Deciding

In this case, 'thinking' meant preparing the next move. There was no new decision to be made because the sequence was fixed. But there are other situations in which thinking means deciding what the next move should be.

In the scanner the choice is again often reduced to choosing which key to press. There is no choice to be made if a light tells you which key to press. But there is a choice if there is no external cue to tell you what to do. This is sometimes described as the 'internally generated' condition because the choice is up to you. The instruction is typically to produce a sequence of key-presses but to avoid long runs of the same press.

In an important study James Rowe and his colleagues showed that the dorsal prefrontal cortex was strongly activated in this condition. Furthermore, the activation was greater the more switches between keys there were in the sequence. But critically it was not present on the first trial on which, by definition, there is no sequence and thus no possibility of a switch. So the activation relates to the attentive control that is needed to produce flexible behaviour.

Waiting to act

In real life, it may not be possible to act on a decision immediately and so it is necessary to put it on hold. Psychologists often use the term 'working memory' for keeping things temporarily in mind, and it is widely believed that the prefrontal cortex is necessary for doing so. However, the fact that something is widely believed often means that it is incorrect.

The problem is the proposal is usually taken to be that the prefrontal cortex is simply involved in 'retention' of what a person has seen or done. This should properly be called 'short-term memory'. Jean Baptiste Pochon and Bruno Dubois assessed this in the scanner by presenting a sequence of lights in different locations. They then tested retention by seeing whether the people recognized the sequence after a delay of six seconds. They found that there was continuous activation throughout the delay in the parietal cortex but not in the prefrontal cortex.

However, the experiment included another condition in which memory was tested by asking the people to repeat the sequence;

and in this condition the dorsal prefrontal cortex was activated during the delay. So the activation in the prefrontal cortex reflected preparation to perform the sequence. In the interests of mental cleanliness it is helpful to distinguish activity that reflects retention (retrospective) from activity that reflects preparation (prospective). The key to the prefrontal cortex is that it is directed towards the future, as in generating sequences or preparing to repeat them.

Planning

There is a difference between waiting before acting and planning. The term planning implies working out what to do. In life we are often faced with tasks that require formulating a plan before having a go. The TV has failed to come on and it helps to think through what might have happened before tinkering.

The ability to plan can be assessed in the laboratory on a computerized version of the 'Tower of London task'. Here the moves are not key-presses but the transfer of billiard balls from one pocket to another. The aim is to work out the minimum number of moves that are needed to rearrange the billiard balls by moving them one at a time from the start array (Figure 18, top) until they match the goal array (Figure 18, bottom). The problem on the left can be solved in two moves but the problem on the right is more difficult. It needs a minimum of five moves and some of them are counter-intuitive.

James Rowe and Adrian Owen have scanned people while they solve problems of this sort by thinking them through in their head. There was activation in two areas, the dorsal prefrontal cortex and the cortex in the intraparietal sulcus, and this matches the finding when people solve other problems that involve reasoning (Chapter 5). However, though these two areas are anatomically connected, they contributed in different ways. The activation in the prefrontal cortex was associated with the

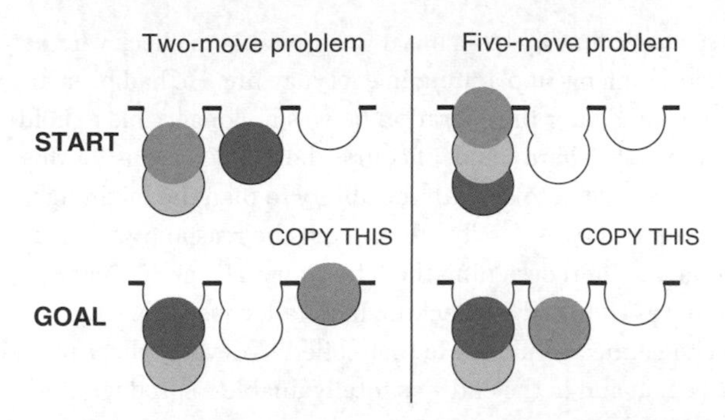

18. Computerized version of the Tower of London. The starting position is shown at the top and the goal position at the bottom. The problem on the left can be completed in two moves and the problem on the right in five moves.

generation of the sequence of moves, whereas the activation in the cortex in the intraparietal sulcus was associated with checking them against the goal array.

So these simple laboratory experiments suggest that the dorsal prefrontal cortex is critically involved in preparing, deciding, and planning for the future. But it is a long way from pressing keys or moving balls to the complexity of everyday life. Fortunately we can close the gap because there are patients with damage to the prefrontal cortex and this means that it is possible to see how they cope with that complexity.

Vinod Goel and Jordan Grafman have described a patient called PF who had a tumour removed from his right frontal lobe. First they checked whether he could do the Tower of London task and it turned out that he could. But he was intelligent and this task is relatively simple. He did poorly on the Tower of Hanoi task from which it was adapted and the reason is that more thought is required on this task because there are more objects to be rearranged.

These are laboratory tasks but it was possible to check whether PF had problems in planning in everyday life. He had been an architect and after the operation he was no longer able to hold down the job. There could, of course, be many reasons for this that have nothing to do with the ability to plan; he might have become lazy. It was possible to find out the reason by assigning him the task of redesigning the laboratory. His preliminary design was confused and lacking in detail, consisting simply of three unconnected ideas. He then failed to develop the proposals that he had and in the end was totally unable to produce a detailed plan.

So PF was unable to generate coherent ideas or put them into practice; and this confirms that the prefrontal cortex is indeed critical for the ability to plan. The evolutionary advantage of being able to try out ideas in the head is that it is possible to imagine the potential outcomes and thus avoid those that are deleterious. One way of putting it is to say that human beings are capable of 'mental trial and error'. It pays to try things out in the head rather than in practice because it avoids the error.

Imagining outcomes

Engaging in mental trial and error demands the ability to imagine the outcomes. In life these outcomes can be fame or esteem, but in the laboratory it is easier to work with money or simple drinks. Chapter 1 has already mentioned the fact that there is activation in the ventromedial prefrontal cortex when people are presented with their favoured drink.

Todd Hare asked people what juice they preferred and then scanned them while they imagined either their preferred juice or water. The finding was that there was activation in the ventromedial prefrontal cortex even when they imagined the juice. However, while it is true that people could imagine their preferred juice, the activation in the primary area for taste was greater when they actually had it in their mouth. When simply thinking of the juice, it is not so easy to re-experience its taste.

There is an important consequence of the fact that our imagination is less detailed. This is that we tend to overvalue outcomes that are immediate. In the same experiment the people were also given the choice between twenty-five dollars now and up to fifty-four dollars at a later date. There was a relation between the choice and the ability to imagine a future reward. The people who chose the immediate option were also those who showed less activation in the ventromedial prefrontal cortex when imagining juice in the future.

Just as it is easy to underestimate future rewards, so it is easy to underestimate future costs. There are two reasons: they are uncertain and they are less vivid than an immediate loss. So we gamble. The propensity for taking a gamble can be assessed on the Iowa Gambling task. In this, the person is shown four packs of cards on a computer screen and is given the choice of turning over the cards one at a time. Some cards give points whereas there are others that deduct points; the number of points accumulated determines the amount of money won by the end of the task.

The choice between packs is tricky because the cards in one of the packs sometimes award big gains, though at other times they signal big losses. It is tempting to choose from this pack but overall the person loses by doing so. The result is that over time most people learn to avoid this pack and turn over cards from the others.

Antoine Bechara and Antonio Damasio have given this task to patients with lesions that include the ventromedial prefrontal cortex and the patients were poor at learning to avoid the risky pack. However, there are two problems in interpreting this result. One is that the task is complex and so failure could occur for several reasons. The other is that in the patients the damage was extensive, including the frontal pole and other parts of the orbital frontal cortex, and so it is not clear which area is critical.

The first problem can be overcome by restricting the analysis to particular trial types; the second by scanning healthy people since the scans should not show up areas that are irrelevant for performance. One such experiment compared the activations when people chose from the risky as opposed to the safe packs. Activation in the ventromedial prefrontal cortex was associated with the risky choice; and the greater the activation, the greater the success in learning to avoid the risky pack.

Why should this be so? The answer has come from studies which checked what happens just *before* a person chooses a card from the risky pack. There is a simple measure of the degree to which the person is alert or anxious. This is the change in sweat on the palm of the hand. This can be measured by the alteration in the electrical resistance, the so-called 'skin conductance response' (SCR).

On the Iowa Gambling task, there is an SCR just before people choose from the risky pack and this suggests heightened alertness. This anticipatory SCR is either reduced or not present in patients with a lesion that includes the ventromedial prefrontal cortex. The implication is that the patients are no longer imagining the potential outcomes when they choose from the risky pack.

The obvious next step is to see whether pathological gamblers show an SCR before choosing from the risky pack, and the answer is that the response is reduced compared with non-gamblers. And as expected they also take more risks on the task. They overvalue immediate rewards despite the danger of long-term losses.

Imagining outcomes for others

In most Western societies whether you gamble is up to you. It may be frowned upon, but it is not regarded as an issue for morality unless your gambling affects your family or others. Moral rules are imposed when it is the good order of society that is at stake.

The fundamental basis for these rules is 'do as you would be done by'. There are, of course, more technical formulations in moral philosophy. But acting on this principle—however phrased—requires you to be able to see things from the other person's perspective, and this involves imagining the effect that your actions would have on that person.

It is one thing to *know* what others might think and another to be able to *feel* empathy with them. To make this comparison, a team of researchers in Sarah-Jayne Blakemore's laboratory studied boys between the ages of eleven and sixteen. The boys were presented with a series of cartoons in which there were two people and they were required to pick one of two possible endings to complete the story.

In the critical condition, the correct choice depended on understanding the feelings of the people: for instance the mother should react by comforting the child when it falls off the slide. In a second condition, the correct choice depended on understanding the thoughts of the people: for instance one person should supply the ladder so that the others can reach the apples on the tree. In a control condition the choice depended on understanding physical causation: for instance the snowman will melt when the sun comes out.

As expected there was activation in the ventromedial prefrontal cortex when the comparison was between the stories with emotional content as opposed to physical causation. There was an indication of a similar effect when the comparison was with stories concerning thoughts, but this effect was both smaller and in need of replication.

So the ventromedial cortex is activated when people think of how others would feel about particular emotional situations. And this is the same area that is activated when someone thinks of juice for themselves or of the risk of losing money. When the boys judged how the mother should react when the child is in trouble, the basis for that judgement involved imagining how the mother would feel. And it is on the ability to feel for others that moral rules depend.

If this is the case, we would predict that damage to the ventromedial prefrontal cortex in early childhood should severely impair the ability to learn these rules. Stephen Anderson and Antonio Damasio have described the case histories of two patients with very early damage that includes this area. The consequences were indeed grave. They stole, lied, and were impervious to punishment; and in general they showed a lack of guilt concerning their actions. You can hear the words that a judge would use in court.

One might expect the result of damage in adults to be different since by then they have learned the rules. However, damage still has a measurable effect. A study in the same laboratory has shown that patients with damage that included the ventromedial prefrontal cortex were impaired in their ability to reason when faced with moral dilemmas. One such was: 'how appropriate is it for you to flip the switch to kill your daughter to save the five workers?' The patients were more likely than healthy people to make utilitarian judgements, seeing no reason not to flip the switch.

There could be two explanations for this and both may apply. One is that they may have been unable to anticipate the feelings of guilt or remorse that they would have if they flipped the switch. The other is that they may have been poor at empathizing with the consequences for the daughter. We know that patients with lesions that include the ventromedial prefrontal cortex fail to show empathy; they do not recognize what it would be inappropriate to say or do given the feelings of others.

Yet both propriety and moral reasoning depend on this recognition. Of course, the fact that most people can feel for others cannot be used to prove that a moral principle must hold. It is a truism in moral philosophy that it is not possible to base a judgement of what 'ought' to happen on anything that 'is' true. In other words no moral principle follows logically because of some fact. Nonetheless, it is not an irrelevant fact that we can see things from the perspective of others. Though it may not justify moral principles, it does mean that we can obey them.

Answers

1. The prefrontal cortex is in a unique position, having access to information about the present situation, information about current needs and aims, and access to the motor system. On this basis it generates the action that is appropriate for the present context. However, where the same situation repeats, actions can become habitual. The penalty is that if the situation suddenly changes, the habitual action may be run off inappropriately. We say that we were 'absent minded'.
2. We are able to plan in our head and imagine potential outcomes. This ability depends on the prefrontal cortex. However, the representation of a future outcome is less vivid than that of an immediate one. The result is that we tend to choose smaller immediate rewards rather than larger but less immediate ones.
3. The moral principle 'do as you would be done by' depends on the ability to intuit what it is like for others to be affected by our actions. The same area is engaged both when we imagine potential outcomes for ourselves and also when we think about how others would feel in a similar situation. This doesn't justify the principle but it does make it possible for us to abide by it.

Chapter 7
Checking

Questions

1. In what sense can my voluntary actions be said to be free?
2. How can we learn from our mistakes when we perform tasks that are difficult?
3. How is it that human beings are able to infer the intentions of others?

Having made a decision, we need to check that we carry it out as intended. In other words we need to monitor our own behaviour.

We live in a world in which management is obsessed with monitoring and feedback. We buy online and the next day we receive an email asking us to rate the site. We buy a mobile phone and before we leave the shop we are asked to rate the assistant.

But we also act as our own managers. We check whether we finished the crossword today as fast as usual. We agitate about how our presentation went. And we keep track of our golf swing to see if it is showing signs of improvement. To put it another way, we are self-regulating.

It is easy to misinterpret this claim. It can be taken to suggest that 'I' tell my brain what to do. Though we are subjectively aware of our thoughts and decisions, it is the brain that monitors and regulates itself, not some other ghostly presence. It is simply that we have an insider's view.

The term 'view' is, of course, metaphorical. It may seem that when I scan the world, I am watching a film outside. And it may seem that when I monitor my decisions, I am watching dials and twiddling the knobs. Sadly metaphors don't make for good science.

Awareness of our intentions

Introspection tells us that when we deliberate we are aware of doing so, but it is difficult to base a scientific account on subjective report. Fortunately, just as Larry Weiskrantz opened up the study of perceptual awareness (Chapter 2), so Benjamin Libet has done for awareness of our intentions.

He did so in one of the great experiments in cognitive neuroscience, published in 1983. The individuals involved were instructed to move their forefinger whenever they wanted; so during the experiment they made a series of movements, spaced at random intervals. At the same time they could see a clock with a spot that moved rapidly around the face; we would now use the screen of a desktop computer for this display (Figure 19). The people were instructed that, when they were aware of their intention to move, they should remember where the spot was on the clock face. Later, after the movement, they were asked to report what that position was; and this told Libet the time at which they were first aware of their intention.

This has come to be known as 'the Libet task'. The result of the original experiment has caused consternation. The reason is that the underlying brain activity started between 300 and 700 milliseconds

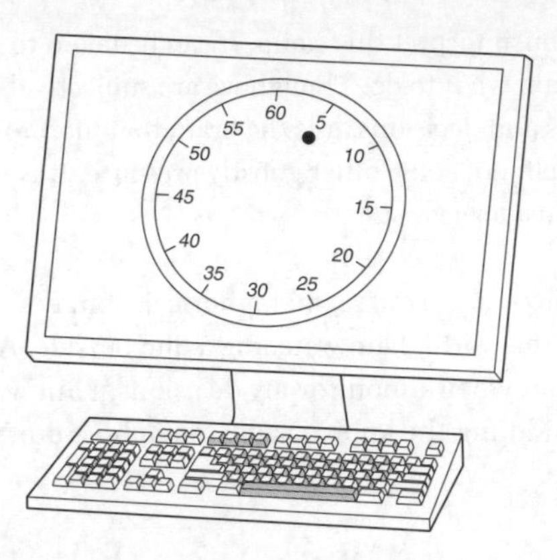

19. Libet clock.

before the people were aware of their decision to move their finger. This activity was measured by placing electrodes on the scalp and recording the 'readiness potential'. This is an electrical event that occurs before spontaneous actions.

Because this potential is measured at the scalp, it is difficult to be certain as to the exact source in the brain of the underlying activity. So Hakwan Lau scanned people with fMRI while they did the Libet task. They either had to time when they were first aware of their intention to move or when they were aware of actually moving their finger. When people timed their intention as opposed to their movement there was an enhancement of the activation on the border between the pre-supplementary motor cortex (Pre-SMA) and supplementary motor cortex (SMA) (Figure 20). The readiness potential reflects preparatory activity in these areas.

The sceptic could argue that this activity may only reflect general preparation, not a specific decision as to *what* to do. So John-Dylan Haynes and his colleagues required people to decide not only when to move but also *which* of two fingers to move.

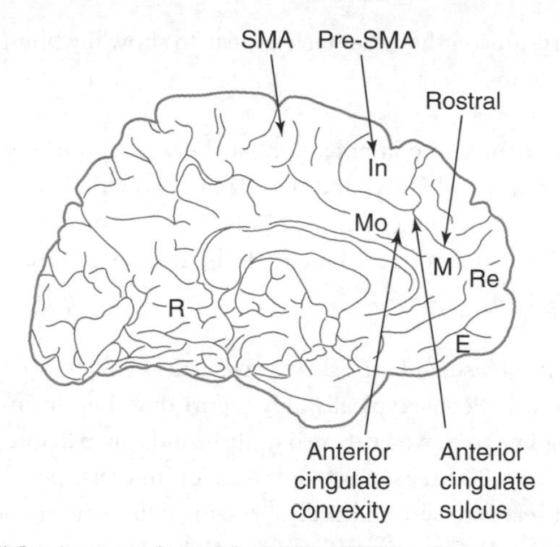

20. Medial or inner surface of the left hemisphere showing areas involved in monitoring. SMA = supplementary motor area, Pre-SMA = pre-supplementary area, the convexity cortex, sulcal cortex, and rostral division of the anterior cingulate cortex. In = intention, Mo = monitoring, M = mentalizing, R = remembering, E = empathizing, Re = reflecting.

The intervals between the movements varied but were of the order of fifteen seconds.

The activation of the prefrontal cortex was studied with a multi-voxel pattern analysis. It turned out that it was possible to predict better than chance which of the two movements the person was going to make, and to do this seconds in advance. This was well before the people were aware of their decision. The sceptic is or should be floored.

So before a voluntary action there is brain activity of which we are unaware. This certainly poses a problem for those philosophers who take the dualist position as advocated by Descartes. Dualism must hold that awareness occurs either *before* the brain activity or at the very least *at the same time*. For a dualist an action can only be said to be free if *I* will it, not my brain. The problem is that the results on the Libet task appear to show my brain dictating to me.

But for someone who is not a dualist there is no 'me' apart from my brain and body; and if so, there is no 'me' for the brain to push about. So long as the brain is healthy, the actions that the brain generates are taken to be the actions that the person generates.

Even so, it is reasonable to ask why awareness comes so late on the Libet task. People typically only report that they are aware that they are going to move 200 or so milliseconds or so before they actually do so. The reason is that the task is one that people can carry out relatively automatically, pressing a finger every now and again. And in the experiment by John-Dylan Haynes, the people were deliberately extensively pre-trained. So when asked, they reported they did the task without thinking.

On theoretical grounds one might expect that the less the attention required, the later people would report awareness. The logic is illustrated in Figure 21, which makes the point by showing two readiness potentials. Suppose that the upper one is for person A and the lower one for person B. If there is a critical threshold for awareness, then person A will be aware earlier than person B, the reason being the greater enhancement of the potential in person A.

Figure 21 illustrates the idea but the data are made up. So in an fMRI study, Hakwan Lau measured the degree of enhancement in the Pre-SMA and related it to the time at which the people reported awareness of their intention. And indeed, the greater the enhancement, the earlier the report.

Nonetheless, there is no escaping the fact that, as suggested in Figure 21, there is subliminal activity in the brain that precedes that awareness. And it is of no help to try to rescue the pride of the

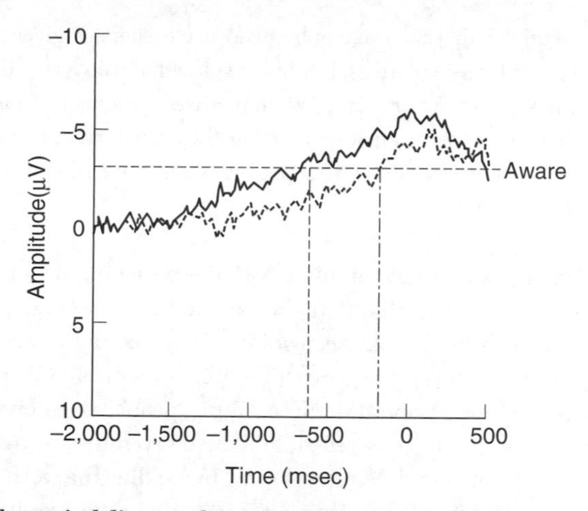

21. Theoretical diagram showing two potentials, one enhanced. The diagram suggests that the signal needs to be of a particular amplitude to reach the threshold for awareness. The diagram also suggests that if the signal is enhanced, awareness will occur earlier.

human race by saying that we can veto that activity. Yes, but there will be subliminal activity before that veto.

The implication is that it is a mistake to define voluntary action in terms that imply early awareness. However late it occurs, awareness still marks the fact that the action is mine. It does not matter that skilled pianists have limited awareness of preparing each and every finger movement. That awareness is enough for them to know that they are moving their fingers at will.

Monitoring responses

The argument is not that we cannot exert a veto on what we were going to do. We can. It is just that it is wrong to characterize the situation as 'me' vetoing what my brain was planning. It is wrong to mix up a psychological account with a neural account of what is happening.

There is indeed an advantage in being aware earlier—it gives more time to consider alternatives, inhibit or veto what we were going to do, and so change our mind. When pianists are learning a new piece, they play it through slowly so that they can correct errors before they occur. The slower the performance, the easier it is to attend to playing the different notes.

Monitoring performance in this way is of especial importance when the task is difficult and there is a danger of making mistakes. One way this has been studied in the laboratory is to use tasks in which it is necessary to inhibit a predominant response. On the 'Stroop task', for example, people are shown words for colours, such as 'red' or 'blue'. However, the words are themselves printed in different coloured fonts: thus the word 'red' might be printed in yellow. Though it is easy to read the word itself ('red'), it is difficult to name the colour of the font ('yellow'). The reason is that there is a conflict and the temptation is to simply say 'red'.

If people are scanned while they are tested on the Stroop task, it is a general finding that there is activation in the anterior cingulate cortex (Figure 20) when the people have to name the colour of the font. But Cameron Carter and Jonathan Cohen have also shown that this activation is especially marked when the task on the previous trial was to name the word.

People are frequently aware when they have made a mistake. In one study people were instructed to press a button when they knew that they had done so. The activation in the anterior cingulate cortex was very much greater on trials on which they detected their error.

The advantage of realizing that you have made a mistake is that you can take especial care on the next trial. So John Kerns and Jonathan Cohen looked to see if they could find evidence for this adjustment. They found that the greater the activation in the anterior cingulate, the greater the degree to which the dorsal prefrontal cortex was re-engaged on the next trial.

If the activation in the anterior cingulate cortex is necessary for this adjustment, it follows that patients with lesions in this area ought to fail to adjust the way in which they treat the next trial. So Giuseppe di Pellegrino and his colleagues recruited patients with extensive lesions that included the anterior cingulate cortex. Rather than testing the patients on the Stroop task, they were assessed on another task in which there are conflicting cues. This is the so-called 'flanker task'. The task is to press the button on the right or left depending on the direction in which a central arrow points. The difficulty is that there are two arrows on either side of the central arrow, and on some trials these point the other way. People are prone to errors on these trials because the signals are confusing.

When people with no lesion made a mistake, they were less likely to do so on the next trial. That is to say, they adjusted as expected. This effect was abolished in the patients with anterior cingulate lesions who failed to learn from their errors.

Monitoring the intentions of others

It is because we are aware of own intentions that we can intuit the intentions of other people. We live in a complex society and harmonious social relations depend on being able to interpret the actions of others. But since we cannot directly observe what they are thinking, we have to infer their intentions from their actions.

In many cases these intentions will be well meaning but there is always the danger that a person will try to deceive, so gaining a competitive advantage. So Julie Grèzes thought up an ingenious experiment on deceit. She prepared videos in which an actor picked up a box. However, on some trials the actor did so in such a way as to make it appear that the box was either lighter or heavier than it really was. People were then scanned while watching these videos and their task was to judge whether the person was trying to deceive them or not.

There were activations in two areas, the temporo-parietal junction (TPJ) (Figure 7, p. 29) and the 'rostral' part of the anterior cingulate cortex (Figure 20). Again anatomists are having a field day. In Latin *rostrum* means 'beak' and so 'rostral' is used here to refer to the most anterior part of the cingulate cortex.

The TPJ is known to send connections to the rostral cingulate cortex. However, in the study by Julie Grèzes there was a difference in the pattern of activation in the two areas. The activation in the TPJ was related to whether the movements were unexpected, irrespective of whether the observers were correct in their judgements concerning deceit. The reason is that the TPJ receives information about motion, and in particular about what is called 'biological motion'; this is the coherent motion of the limbs and body.

By contrast, the activation in the rostral cingulate cortex was related to the accuracy of the judgements about whether the actor was or was not trying to deceive. Here the activation occurred when the observers were correct about the intentions of the actor.

It is not necessary to present videos. In a delightful study Chris Frith showed people cartoons. Half of them were funny, but it was only possible to get the joke if you could read the intentions of the people depicted in the cartoon. Chris Frith illustrates this with a cartoon in which there is a man sitting on a ledge and fishing in the sea. Above him, another man sits on the top of the cliff with his line dangling in the basket of the man lower down the cliff. You only get the point if you can appreciate that the man at the top of the cliff is trying to steal the fish caught by the man below. When people were scanned while viewing cartoons of this sort, there was activation in the rostral cingulate cortex.

Mentalizing

The ugly word 'mentalize' has been used to refer to the ability to infer the thoughts and intentions of others. There have been two suggestions about how we do it. One is that we simulate the actions ourselves, and then read off the intentions automatically. This suggestion is based on work on animals in which neurons have been found that show a burst of activity both when the animal performs a particular action and when it sees another animal performing the same action. These have been called 'mirror neurons'.

The alternative is that we learn to predict what people are going to do by relying on the fact that we have long experience in people watching. So this could provide a knowledge base of human nature. We could use this to interpret the actions of others.

To find out which of these alternative accounts is correct, a group in Ivan Toni's laboratory showed people a series of photos. Some showed ordinary actions, such as drinking from a cup. Others showed extraordinary ways of carrying out the actions, for example drinking from the cup while holding it by its mouth. Yet others showed the person acting in a way that was inexplicable given ordinary intentions, such as holding the cup to their hair.

The observers were scanned while they did two tasks. In one they were instructed to attend to the intention of the actor and in the other to attend to the means being used by the actor. When the observers saw the pictures showing extraordinary means, there was activation near the TPJ. This activation occurred when viewing photos in which movements were odd, just as in the study by Julie Grèzes using videos.

When the observers viewed the photos showing extraordinary intentions, there was activation in the ventral prefrontal cortex. This is one of the areas in which mirror neurons have been found in animals. So this activation was taken to suggest covert simulation.

However, there was also an effect that depended on whether the observers were attending to the intention of the actor as opposed to the means used. There was activation in the rostral cingulate cortex when the observers attended to the intention. This activation was taken to suggest reflecting on mental states.

The implication is that interpreting the intentions of others depends on two processes. One was discussed in Chapter 6: that we intuit the intentions or feelings of others by reading off our own. The other is that we have a lifetime's experience of observing others and thus the opportunity to build up an extensive knowledge of how other people act.

There is a way of proving that knowledge plays a part and this is to scan people while they make judgements that can only be made on the basis of knowledge. For example, Kevin Ochsner asked people to choose from a set of adjectives applied to themselves or those that applied to other people that they knew. The adjectives related to traits, as in 'kind' or 'angry'. There were activations in the rostral cingulate cortex irrespective of whether the people were thinking about themselves or others, and the peaks were in very similar locations for the two conditions. It is, of course, a matter of knowledge, not introspection, whether adjectives are appropriate for other people with whom we are familiar.

Metacognition

The ability to reflect on the intentions of others involves what is called 'metacognition'. This means 'thinking about thinking'.

We do not know to what extent our nearest primate relatives are capable of doing this but we can be certain that the ability is greatly enhanced in human beings.

To understand how this has come about, we need to understand the way in which the medial surface is organized. The key to the medial surface is that it concerns the self. As mentioned in Chapter 4, there are widespread activations on the medial surface including the hippocampus when we remember events that are personal to ourselves (Figure 20, 'R' for remembering).

There is also a posterior to anterior organization on the medial surface. There is activation near the Pre-SMA when we attend to our intentions (Figure 20, 'In' for intentions) and in the anterior cingulate cortex when we monitor our responses (Figure 20, 'Mo' for monitoring). But the rostral cingulate cortex receives an input that relates to the movements of others and it is for this reason that the activations are yet more anterior when we mentalize (Figure 20, 'M' for mentalizing) or empathize with the feelings of others (Figure 20, 'E' for empathizing). The medial prefrontal cortex lies at the front, and this is activated when we reflect on events, whether in the past or future (Figure 20, 'Re' for reflection).

Chapter 5 has already mentioned that the prefrontal cortex, including the medial prefrontal cortex, is especially enlarged in the human brain. But the anterior cingulate cortex, including the rostral cingulate cortex, is also greatly expanded compared with the brain of a chimpanzee. Since metacognition involves representing a representation (thinking about thinking), one way of achieving this could be via a hierarchical organization. As explained in Chapter 2, this type of organization allows the re-representation or categorization of earlier inputs. So the increase in size of the human anterior cingulate cortex may reflect the addition of more layers to the hierarchy.

Answers

1. If we are asked to move a finger whenever we want, it is possible to detect activity in the brain that precedes the time at which we are aware that we have made the decision to move. This is only problematical if one takes the dualist position in philosophy. If one rejects this position, then there is no 'me' for the brain to push about. The fact that awareness follows subliminal activity in the brain does not change the fact that my actions are those that are generated by my brain and that they are voluntary so long as no one else has forced me to act in that way.
2. Given that we are aware of what we intend to do, we can then compare it to what we actually do. When we are performing a difficult task and we make an error, the prefrontal cortex is re-engaged on the next trial. This means that we attend better on the next trial and are thus less likely to make an error.
3. We are not only aware of our own mental states, we also have the ability to infer the mental states of others. When we observe the action of someone else, we automatically simulate that action and can thus intuit their intention. But we are also able to apply our past experience and knowledge so as to reflect on what they are most likely to be thinking. There is overlap between cortex that is engaged when we think about ourselves and when we think about others.

Chapter 8
Acting

Questions

1. Why is there a relation between handedness and the cerebral hemisphere that is specialized for language?
2. How is it that we can learn to play the violin?
3. Why can't we tickle ourselves?

Like other primates we interact with the world by using our hands. However, we differ in three ways. One is the facility with which we are able to move each finger independently. It is this that makes it possible for musicians to cope with their instruments, whether the saxophone, the guitar, or the keyboard.

Another difference is that the majority of people prefer to use their right hand for skilled actions such as writing or throwing. Even though individual chimpanzees may show a preference for one hand for a particular task, populations of chimpanzees in the wild show no tendency towards right-handedness.

But the overriding difference is that there is a relation between handedness and speech. In right-handers it is lesions in the left

Broca's area that impair the production of spoken language. And it is lesions in the left Wernicke's area that impair the comprehension of spoken language.

So we need to explain why there should be a relation between handedness and hemisphere dominance for speech.

Right-handedness

The archaeological record suggests that the tendency to right-handedness goes back as far as 1.5 million years. The evidence comes from studying stone tools. The flakes are struck from the stone core with the rapid movement of one hand, and the marks on the core differ according to the direction from which the strike comes.

There is a clear advantage for the individual in training up one hand in this skill. Since it takes long practice to acquire, using the same hand means that it is the same hemisphere that learns. This will be the hemisphere that is opposite to that hand.

But this does not explain why different individuals should be alike in their hand preference. Initially it could be that different individuals had similar instruction in the art of stone knapping, or it could be that different individuals used the same cores to strike off flakes. Either way there would be a tendency towards increasing standardization over the generations.

Left hemisphere specialization for gestures

It looks as if handedness came before spoken language. Our best estimate is that speech evolved between 70,000 and 100,000 years ago. But many have speculated that hominins might well have used a gestural system for communication before they developed spoken language. If so, the left hemisphere would be involved because of the tendency to right-handedness. In that case a specialization for gestural communication would have formed a 'pre-adaptation' for spoken language.

There is evidence in favour of the speculation. This is that we know that in modern humans the left hemisphere is critical for gestural communication. Ursula Bellugi has studied people who use American Sign Language to communicate because they are deaf; and she has reported that left but not right hemisphere lesions severely impaired their ability to produce and understand gestures. This is not because the patients had motor or sensory loss: when they named objects they often produced the wrong gesture, in other words the one that was appropriate for another object.

Karen Emmorey has also scanned hearing people with PET while they named things using American Sign Language or speech, and she found activation in the left Broca's area in both cases. However, there was a significant difference when the people were tested on their ability to understand signs or spoken words. This is that there was activation in the left inferior parietal cortex when they observed the gestures. The reason is that it receives information about biological motion (Chapter 5). When people judge whether two gestures are the same or different, the activation is in the left inferior parietal cortex.

It is therefore of especial interest that the anterior part of the inferior parietal cortex connects directly to Broca's area (Figure 22). And when the fibre tract is visualized using diffusion weighted imaging it turns out that it is more extensive in the left than the right hemisphere. This asymmetry is not present in the brain of chimpanzees.

In the human brain there are also connections to Broca's area from Wernicke's area in the superior and middle temporal gyrus (Figure 22). And again in the human but not chimpanzee brain the connections are asymmetrical, favouring the left hemisphere.

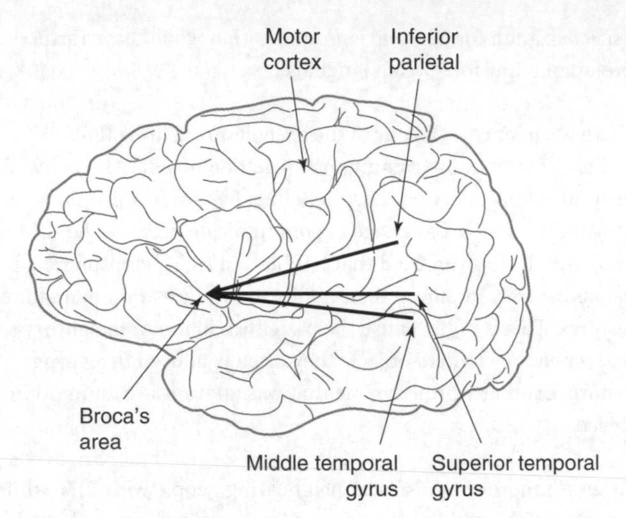

22. The connections of Broca's area.

So in the human brain Broca's area has access to information about gestures as well as information about the sounds of speech. Unfortunately no patients have been described with lesions that are strictly limited to Broca's area; we would expect them to have problems with both gestural and spoken language. However, Gereon Fink and his colleagues have compared the lesions of a group of fifty patients who had suffered left hemisphere strokes. Of these, 75 per cent had problems with language; in neurological terminology they were 'aphasic'. And 47 per cent of the patients had problems with imitating gestures; in neurological terminology they were 'apraxic'. Neurologists like their Greek.

There were differences between the lesions for the two groups. For example, the patients who were poor at using gestures were more likely to have lesions in the inferior parietal cortex and ventral prefrontal cortex. But the two groups were similar in having damage to Broca's area itself. The suggestion is that Broca's area forms a common hub for the two systems, gestural and vocal.

So which came first, gesture or speech? It is difficult to see why there should be an asymmetry in the connections in the gestural system *as the result of* spoken language. So it seems more likely that gestural communication was the prime mover.

Spoken language

Spoken language has an advantage over a gestural system. Names can be given to objects in either system, but it is much easier to generate an infinite variety of possible sentences in spoken language. The linguist Noam Chomsky has long argued that these are not the result of simply stringing words together in sequences, but that there is a hierarchical system for generating grammatical sentences. And he insists that this mechanism evolved *de novo*, meaning that it could not have been the result of the slow adaptation of prior mechanisms for producing sequences. This is a challenging claim. It is on a par with claiming—wrongly—that the mammalian eye could not have evolved in a series of stages, all of which were adaptive.

So there are two fundamental questions. One is whether it is true that the human brain is specialized for learning hierarchical organized sequences. The other is how it acquired that specialization.

Angela Friederici and her colleagues have used brain imaging to address the first question. They developed two types of artificial grammar. In one the rules simply related to local dependencies, as in A1, B1, A2, B2. The other generated sequences hierarchically, as in A2, A1, B1, B2 in which there is a central nested structure. Both systems used syllables rather than words. People were taught until they were proficient in both grammars and were then scanned while they were presented with the syllables on the screen and had to judge whether the sequences were or were not grammatical. There were many areas in which the activation was greater for the hierarchical system but the authors lay stress on the fact that this was also true for Broca's area on the left.

However, there is an important difference between the two grammars. This is that in the hierarchical system there are long distance dependencies, and the longer the phrase structure the greater the need to hold earlier words in temporary memory. So it is not clear whether the enhancement of the activation in Broca's area reflects memory for the syllables or the association of the syllables over a distance.

So Karl Magnus Petersson and Peter Hagoort compared artificial grammars that were matched for the familiarity of strings that were adjacent or non-adjacent. When people were scanned while they made judgements about the grammaticality of hierarchical sequences, there was activation in the region of Broca's area irrespective of whether the judgements were about dependencies that were local or distant.

But this does not answer the question about how the mechanisms for learning such sequences could have evolved. The question is best answered by devising computer models that are capable of learning artificial grammars and seeing to what extent their properties have to be specific to that task. Karl Magnus Petersson and Peter Hagoort have developed a successful neurobiological model that incorporates mechanisms for forming and storing associations over distance. The properties of the model are general enough that it can learn a wide variety of tasks that have in common the need for processing structured sequences. This suggests that there is no need to suppose that in evolution the mechanisms for syntax evolved *de novo*.

Skill

However language is structured and however it evolved, children need to learn to speak it and to do so automatically. This is a long and difficult process. And there are many other skills that the child has to learn: how to pick objects up, how to stand and walk, and in Japan how to play the violin by the Suzuki method.

One way to investigate how skills are learned is to teach new ones to adults. The nature of skill in general is that it involves precision whether in the application of force, in the control of rhythm, or in the coordination of different movements. All these are involved in learning the violin.

In a study of skill, Anna Floyer-Lea and Paul Matthews taught people to apply a sequence of forces with their forefinger so as to continuously track the movement of a bar on the screen. As the sequence became automatic, the people no longer needed to rely on the visual display because they had learned that there was a repeating sequence. This meant that one force acted as a cue for the next one; and the finding was that as the sequence was learned there was a marked increase in activation in the cerebellum which lies at the base of the brain (Figure 23).

To show that automatic skills can be performed without reference to external cues, Narender Ramnani taught people to move a finger with a particular rhythm. A visual cue paced the rhythm, but as the people learned, so the activation in visual areas

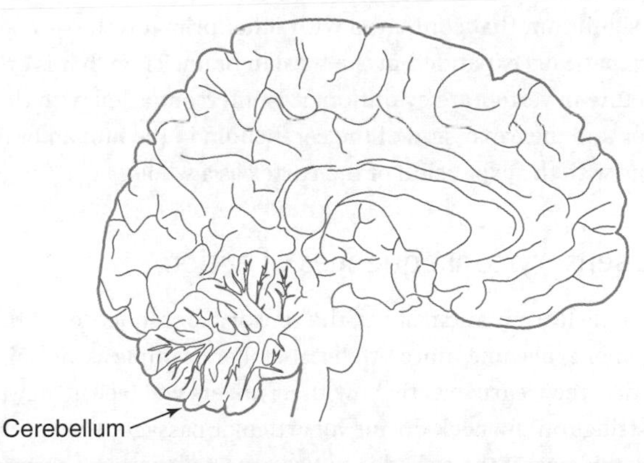

23. Medial or inner surface of the brain with the brain stem and cerebellum.

decreased. In other words, with practice the rhythm could be performed without external cueing. At the same time there was a marked increase in the activation of the cerebellum, as in the previous study.

The cerebellum is also involved in language skills. Marcus Raichle and Steve Petersen scanned people while they were presented with nouns. For each noun they had to produce a verb that was appropriate: for example, given the word 'cake' they might produce the word 'eat'. The same series of nouns was repeated throughout the experiment and this encouraged the people to repeat themselves, for example always saying 'eat' when the noun 'cake' came around. The consequence was that by the end they were producing a repeating sequence of verbs, and this was associated with an increase in the activation of the cerebellum.

Given these results it is not surprising that damage to the cerebellum interferes with both manual and vocal skills. Patients with cerebellar lesions are often clumsy with their hands and poor at articulating with precision. Inevitably, neurologists have a word for poor articulation: 'dysarthria'.

It is significant that compared with other primates the cerebellum is dramatically expanded in the human brain. Though it is an old structure in vertebrate evolution, it is interconnected with the cortex and the expansion of the cerebellum in the human brain is in line with the expansion of the cortex as a whole.

The sensory consequences of action

As people learn manual skills, the sensory consequences of the movements become more predictable. For example as violinists practice, they learn exactly how their fingers will feel as they press the strings on the neck during a particular passage. They are also quick to detect if the sound is not as expected, perhaps because of poor tuning of the strings.

This effect can be studied by using brain imaging. Tim Ebner taught people to move a joystick so as to superimpose the cursor on a target. When they had learned to do this, the relation between the movement of the joystick and the cursor was reversed so that movements to the left caused the cursor to move to the right, downwards movements caused the cursor to move up, and so on. When the people were scanned, there was a signal in the cerebellum when the cursor suddenly moved in an unexpected direction. A signal of this sort is said to mark a 'prediction error'.

There is another way of demonstrating the same effect. This is to make use of what seems at first to be a trivial observation: people find it difficult to tickle themselves. Of course everyone knows this to be true but scientists are unusual in being obsessed with asking *why* things are the case.

Sarah-Jayne Blakemore, Chris Frith, and Dan Wolpert devised a robot arm with a rubber tip such that, when the person moved the arm, the rubber moved on their hand. If the rubber moved as expected, it did not feel at all ticklish. But if the robot arm was fixed such that when the person moved the arm the rubber tip moved with a slight delay or change in direction, the people reported that it tickled.

Given this set-up, people could then be scanned while the degree to which the tip moved out of sync was varied. Though the people did not notice the change, they found it more and more ticklish the greater the delay. And the greater the degree of asynchrony, the more the cerebellum was activated. Again, a signal for a prediction error.

There is a similar signal in the anterior striatum (Figure 17), but this relates not to the sensory consequences but to the outcomes as measured in terms of reward. John O'Doherty taught people to perform actions so as to gain either juice or money as a reward.

When either the juice or the money failed to appear as expected there was a signal marking a prediction error in the striatum.

The notion of a prediction error is fundamental to psychology. In trying to understand the world, we are interested in events that are novel and not as predicted; these indicate that our understanding is not yet complete and so they fire our curiosity. Similarly, as we operate in the world, we are particularly concerned when our actions fail to have the consequences that we expected; the situation calls for a rethink.

Of course, when I was a student we were not encouraged to talk in these terms. Behaviourism banned terms such as 'expect' because they refer to events in the head. But then for the same reason it also banned terms such as 'attending', 'remembering', or 'deciding'. Fortunately, psychology has come a long way since then.

Answers

1. There is a relation between right-handedness and left hemisphere dominance for speech. It is likely that in human evolution the tendency to right-handedness developed before speech. In modern humans, Broca's area is activated both during spoken language and gestural language. Thus, if the hominins used their hands to communicate by gesture, this could have served as a pre-adaptation for the asymmetry for spoken language.

2. Vocal skill requires the ability to move the laryngeal muscles independently and to coordinate the timing with the intake of breath. Similarly, manual skills such as learning the violin require independent movements of the fingers and precise timing. In each case the automation of skills depends on cerebellar mechanisms, and in the human brain the cerebellum has expanded in line with the neocortex.

3. We are less interested in expected than in novel events because these may require changes in our understanding or behaviour. Signals for prediction error can be recorded both in the cerebellum and the anterior striatum, the former when the sensory consequences are not as expected and the latter when the rewarding outcomes are not as expected. The reason why we can't tickle ourselves is that we can predict the sensory consequences when we try. It is when someone else tickles us that the situation is uncomfortable: it is unpredictable. Hence our protest.

Chapter 9
The future

In the mid-1960s, when I trained in clinical psychology, there was only one way in which a brain tumour could be detected in the head. This was by injecting air into the spinal cord so that it reached the ventricles, the fluid-filled cavities in the brain. A simple X-ray then allowed radiologists to visualize the distortion that the tumour caused to the ventricles. The method was crude and it resulted in a dreadful headache.

It was inconceivable at the time that we would be able to take pictures of the brain that are so lifelike—and without the headache. And the idea that we would be able to use multi-voxel pattern analysis to decode the information in an area of the brain would have been thought to be a topic for Aldous Huxley.

So it is tempting to speculate about what technical advances might be possible that could revolutionize cognitive neuroscience. The problem is that speculations concerning the future course of science are almost invariably wrong. Too much is down to chance discovery and serendipity.

We can, however, consider what it is that we need to know and what methods that are currently available or in development might be able to tell us. Now that we know so much about *where* there is activity in the brain while people perform cognitive tasks, the next stage is to find out *how* that activity makes cognition possible. In other words we need to understand the mechanisms.

To do this we need to study that activity with methods that are sensitive to the temporal pattern of neuronal activity. But neither PET nor fMRI is well suited for the purpose because the blood supply does not change as rapidly as the activity of the neurons themselves. This means that we will need to exploit other methods and there are currently three that are available.

Magneto-encephalography

Of these the one that is currently most promising is magneto-encephalography (MEG). This is a method that enables us to detect the minute magnetic signals that are associated with the electrical activity of neurons. Whereas in an MRI scanner the person lies horizontally, in an MEG scanner the person sits upright with their head in the scanner.

MEG has the advantage that it can distinguish the timing of events as they occur throughout the cortex. So rather than simply saying *where* events occur, it can be used to study the *order* of the events as the activity progresses through the cortex. For example, Rita Hari scanned people with MEG while they watched a face with moving lips. The task was to imitate the movements. It was possible to follow the progress of the activity from visual areas to the motor system (Figure 24).

There is current excitement over MEG because there is a serious prospect that it will become portable and it may not even be necessary for the person to be sitting. The advantage would be that the range of behaviour that could be studied would greatly increase.

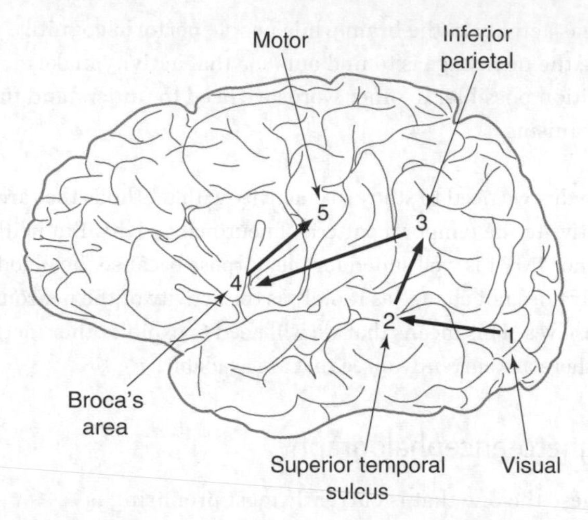

24. Order in which brain areas become active when people imitate mouth movements as recorded with magneto-encephalography.

Recording with electrode arrays

The second method involves implanting arrays of electrodes under the skull. This is done for a week or so in some epileptic patients before they undergo neurosurgery. The advantage is that it is then possible to give tests to the patients and relate aspects of performance to changes in the neuronal activity.

Richard Andersen and his colleagues have used this method to record from the parietal cortex in a patient who is tetraplegic. By analysing the patterns of activity across the electrodes they have been able to tell where the patient is thinking of moving to, even though he is unable actually to do so.

The ultimate aim of research of this sort is to use the recordings to control either robotic arms or legs or actual arms and legs so as to move them in the desired direction. It has already been shown that this is feasible. Recordings were taken from the brain of a patient with tetraplegia and these were used to control the movements of a virtual arm as displayed on a computer screen. The patient learned to move the arm simply by thinking of the desired direction.

Recording from single neurons

Finally, there are situations in which it is judged clinically justifiable to insert fine electrodes into the brain so as to record the activity of single neurons during the surgery itself. The procedure is painless because there are no pain receptors in the brain.

Itzhak Fried and Christof Koch have inserted electrodes into the hippocampus and surrounding areas during surgery, and have recorded while the patients viewed pictures of famous people. They found individual neurons that responded selectively to particular people. Many a distinguished professor has lectured on how one neuron only showed a burst of activity in response to a picture of Jennifer Aniston—though the professor hasn't a clue who on earth Jennifer Aniston actually is.

It is indeed surprising to find neurons that code so selectively but it is easy to be misled. In the first place, the same neuron responded when the patient read the name; so the activity reflects the memory of Jennifer Aniston rather than the perception of her. Secondly, since it was only possible to record from a small sample of the neurons, there is no evidence that this was the only neuron in the region that responded in this way.

Biologically plausible computational models

So what of the future for the longer term? No one believes that all we need to do to understand the brain is to use brain imaging or record from neurons. There are an estimated 100 billion neurons in the human brain. Thus we need models of how the brain works that we can run on supercomputers.

These models will need to take into account two features of the brain. The first is its fine architecture. This has already been done for the area of a mouse somatosensory cortex in which the neurons respond to the bending of a whisker. This achievement took the cooperation of many European laboratories and the massive computing power of IBM's Blue Brain computer.

The second is the brain's gross architecture, meaning the types of structure that are involved. A start on this has been made by the company Deep Mind, now owned by Google. This has gained media attention by developing the programme AlphaGo which has beaten a world champion player of the board game Go. Like the brain, the architecture of the programme includes planning and value networks and on this basis it can learn to predict the expected outcome of any particular position.

It is clear, then, that the long-term aim must be to produce computer models that are more and more flexible and that are more and more biologically plausible in terms of what we know about the actual brain.

So let us suppose that is done.

Will that be psychology or will it be neuroscience? The question is much like asking whether molecular biology is physics, chemistry, or biology. Like molecular biology, cognitive neuroscience lies at the borders between different disciplines, and that is where the excitement so often lies in science. Much more exciting than studying the behaviour of rats and pigeons.

References

Chapter 1: A recent field

Pepsi and Coca Cola:

McClure, S. M. et al., 'Neural correlates of behavioral preference for culturally familiar drinks'. *Neuron*, 2004. 44: 379–87.

Chapter 2: Perceiving

Patient Dee:

Goodale, M. A. and A. D. Milner, *Sight Unseen*. 2013, Oxford: Oxford University Press.

Scanning colour and motion:

Zeki, S. and J. Stutters, 'Functional specialization and generalization for grouping of stimuli based on colour and motion'. *NeuroImage*, 2013. 73: 156–66.

TMS over visual cortex:

Cowey, A. and V. Walsh, 'Tickling the brain: studying visual sensation, perception and cognition by transcranial magnetic stimulation'. *Progress in Brain Research*, 2001. 134: 411–25.

Stimulating the stump in amputees:

Bjorkman, A. et al., 'Phantom digit somatotopy: a functional magnetic resonance imaging study in forearm amputees'. *European Journal of Neuroscience*, 2012. 36: 2098–106.

Scanning synaesthetes:

Nunn, J. A. et al., 'Functional magnetic resonance imaging of synesthesia: activation of V4/V8 by spoken words'. *Nature Neuroscience*, 2002. 5: 371–5.

Connections in syaesthesia:

Rouw, R. and H. S. Scholte, 'Increased structural connectivity in grapheme-color synesthesia'. *Nature Neuroscience*, 2007. 10: 792–7.

Hierarchical processing:

Grill-Spector, K., et al., 'A sequence of object-processing stages revealed by fMRI in the human occipital lobe'. *Human Brain Mapping*, 1998. 6: 316–28.

View independent representation:

Grill-Spector, K., et al., 'Differential processing of objects under various viewing conditions in the human lateral occipital complex'. *Neuron*, 1999. 24: 187–203.

Categorization of animals:

Connolly, A. C., et al., 'The representation of biological classes in the human brain'. *Journal of Neuroscience*, 2012. 32: 2608–18.

Guessing in patient with VI lesion:

Weiskrantz, L., 'Blindsight revisited'. *Current Opinion in Neurobiology*, 1996. 6: 215–20.

Backward masking to compare aware with unaware state:

Dehaene, S. and J. P. Changeux, 'Experimental and theoretical approaches to conscious processing'. *Neuron*, 2011. 70: 200–27.

Scanning with eyes shut and eyes open:

Riedl, V., et al., 'Local activity determines functional connectivity in the resting human brain: a simultaneous FDG-PET/fMRI study'. *Journal of Neuroscience*, 2014. 34: 6260–6.

Saliency network:

Menon, V. and L. Q. Uddin, 'Saliency, switching, attention and control: a network model of insula function'. *Brain Structure Function*, 2010. 214: 655–67.

Effect of propofol on awareness:

Ni Mhuircheartaigh, R., et al., 'Slow-wave activity saturation and thalamocortical isolation during propofol anesthesia in humans'. *Science Translational Medicine*, 2013. 5: e208ra148.

Chapter 3: Attending

Spatial maps in parietal and frontal eye field:

Jerde, T. A., et al., 'Prioritized maps of space in human frontoparietal cortex'. *Journal of Neuroscience*, 2012. 32: 17382–90.

Eye movements in neglect:

Mannan, S. K., et al., 'Revisiting previously searched locations in visual neglect: role of right parietal and frontal lesions in misjudging old locations as new'. *Journal of Cognitive Neuroscience*, 2005. 17: 340–54.

Right hemisphere dominance in the TPJ:

Shulman, G. L., et al., 'Right hemisphere dominance during spatial selective attention and target detection occurs outside the dorsal frontoparietal network'. *Journal of Neuroscience*, 2010. 30: 3640–51.

Multi-voxel analysis of visual search:

Peelen, M. V. and S. Kastner, 'A neural basis for real-world visual search in human occipitotemporal cortex'. *Proceedings National Academy Sciences USA*, 2011. 108: 2125–30.

Auditory memory task as a distraction:

Klemen, J., et al., 'Auditory working memory load impairs visual ventral stream processing: toward a unified model of attentional load'. *Journal of Cognitive Neuroscience*, 2010. 22: 437–46.

Distraction from pain:

Sprenger, C., et al., 'Attention modulates spinal cord responses to pain'. *Current Biology*, 2012. 22: 1019–22.

Effect of verbal instructions:

Hartstra, E., F. Waszak, and M. Brass, 'The implementation of verbal instructions: dissociating motor preparation from the formation of stimulus-response associations'. *NeuroImage*, 2012. 63: 1143–53.

Interactions between prefrontal and task specific areas:

Sakai, K. and R. E. Passingham, 'Prefrontal interactions reflect future task operations'. *Nature Neuroscience*, 2003. 6: 75–81.

TMS of frontal cortex and effects on MT and FFA:

Heinen, K., et al., 'Direct evidence for attention-dependent influences of the frontal eye-fields on feature-responsive visual cortex'. *Cerebral Cortex*, 2013. 24: 2815–21.

Cost of multitasking in terms of time and errors and the effect of automation:

Grol, M., 'Cerebral changes during performance of overlearned arbitrary visuomotor associations'. *Journal of Neuroscience*, 2006. 29: 117–25.

Any input for any output in ventral prefrontal:

Tamber-Rosenau, B. J., et al., 'Amodal processing in human prefrontal cortex'. *Journal of Neuroscience*, 2013. 33: 11573–87.

Cue and response components in ventral prefrontal:

Toni, I., et al., 'Signal-, set- and movement-related activity in the human brain: an event-related fMRI study'. *Cerebral Cortex*, 1999. 9: 35–49.

Effect of automation on the length of the time of activation:

Dux, P. E., et al., 'Isolation of a central bottleneck of information processing with time-resolved FMRI'. *Neuron*, 2006. 52: 1109–20.

Chapter 4: Remembering

Parahippocampal cortex and scenes:

Epstein, R. and N. Kanwisher, 'A cortical representation of the local visual environment'. *Nature*, 1998. 392: 598–601.

Location of individual in scene:

Hassabis, D., et al., 'Decoding neuronal ensembles in the human hippocampus'. *Current Biology*, 2009. 19: 546–54.

Hippocampus activated selectively during navigation:

Spiers, H. J. and E. A. Maguire, 'Thoughts, behaviour, and brain dynamics during navigation in the real world'. *NeuroImage*, 2006. 31: 1826–40.

Taxi driver navigating London after stroke:

Maguire, E. A., R. Nannery, and H. J. Spiers, 'Navigation around London by a taxi driver with bilateral hippocampal lesions'. *Brain*, 2006. 129: 2894–907.

Patient Jon:

Vargha-Khadem, F., et al., 'Differential effects of early hippocampal pathology on episodic and semantic memory'. *Science*, 1997. 277: 376–80.

Activations in hippocampus during recall of events and reflection on future ones:

Addis, D. R. et al., 'Remembering the past and imagining the future: common and distinction neural substrates during event construction and elaboration'. *Journal of Neuroscience*, 2007. 45: 1363–77.

Episodic memory in scrub jays:

Clayton, N. S., 'Can animals recall the past and plan for the future?' *Nature Reviews Neuroscience*, 2003. 4: 685–91.

Episodic memory in chimpanzees:

Sayers, K. and C. R. Menzel, 'Memory and foraging theory: chimpanzee utilization of optimality heuristics in the rank-order recovery of hidden foods'. *Animal Behaviour*, 2012. 84: 795–803.

Reinstatement of memories in people:

Bird, C. M., et al., 'Consolidation of complex events via reinstatement in posterior cingulate cortex'. *Journal of Neuroscience*, 2015. 35: 14426–34.

Reinstatement of memories in mice:

Tanaka, K. Z., et al., 'Cortical representations are reinstated by the hippocampus during memory retrieval'. *Neuron*, 2014. 84: 347–54.

Mind wandering:

Mason, M. F., et al., 'Wandering minds: the default network and stimulus-independent thought'. *Science*, 2007. 315: 393–5.

Medial system in chimpanzees:

Barks, S. K., et al., 'The default mode network in chimpanzees (Pan troglodytes) is similar to that of humans'. *Cerebral Cortex*, 2015. 25: 538–44.

Use of multi-voxel pattern analysis to distinguish recent and remote memories:

Bonnici, H., et al., 'Detecting representations of recent and remote autobiographical memories in vmPFC and hippocampus'. *Journal of Neuroscience*, 2012. 32: 16982–91.

Activations during the retrieval of semantic knowledge:

Vandenberghe, R., et al., 'Functional anatomy of a common semantic system for words and pictures'. *Nature*, 1996. 383: 254–6.

Object encoding and the perirhinal cortex:

Clarke, A. and L. K. Tyler, 'Object-specific semantic coding in human perirhinal cortex'. *Journal of Neuroscience*, 2014. 34: 4766–75.

Spatial disorientation in Alzheimer's disease:

Luzzi, S., et al., 'The neural correlates of road sign knowledge and route learning in semantic dementia and Alzheimer's disease'. *Journal of Neurology Neurosurgery Psychiatry*, 2015. 86: 595–602.

Chapter 5: Reasoning

Scanning Raven's Matrices:

Christoff, K., et al., 'Rostrolateral prefrontal cortex involvement in relational integration during reasoning'. *NeuroImage*, 2001. 14: 1136–49.

General intelligence:

Duncan, J., et al., 'A neural basis for general intelligence'. *Science*, 2000. 289: 457–60.

Effects of lesions on performance on tests of fluid intelligence:

Woolgar, A., et al., 'Fluid intelligence loss linked to restricted regions of damage within frontal and parietal cortex'. *Proceedings National Academy Sciences USA*, 2010. 107: 14899–902.

Rotating objects in the head:

Jordan, K., et al., 'Cortical activations during the mental rotation of different visual objects'. *NeuroImage*, 2001. 13: 143–52.

Scanning knowledge of order:

Fias, W., et al., 'Processing of abstract ordinal knowledge in the horizontal segment of the intraparietal sulcus'. *Journal of Neuroscience*, 2007. 27: 8952–6.

Relation between number and size:

Pinel, P., et al., 'Distributed and overlapping cerebral representations of number, size, and luminance during comparative judgments'. *Neuron*, 2004. 41: 983–93.

Scanning inner speech:

Shergill, S. S., et al., 'Modulation of activity in temporal cortex during generation of inner speech'. *Human Brain Mapping*, 2002. 16: 219–27.

Syllogistic reasoning:

Goel, V., et al., 'Dissociation of mechanisms underlying syllogistic reasoning'. *NeuroImage*, 2000. 12: 504–14.

Parietal lesions and transitive inference:

Waechter, R. L., et al., "Transitive inference reasoning is impaired by focal lesions in parietal cortex rather than rostrolateral prefrontal cortex'. *Neuropsychologia*, 2013. 51: 464–71.

Reasoning in the absence of inner speech:

Levine, D. N., R. Calvanio, and A. Popovics, 'Language in the absence of inner speech'. *Neuropsychology*, 1982. 20: 391–409.

Scanning reasoning about analogies:

Wendelken, C., et al., '"Brain is to thought as stomach is to ??": investigating the role of rostrolateral prefrontal cortex in relational reasoning'. *Journal of Cognitive Neuroscience*, 2008. 20: 682–93.

Multiple demand system:

Fedorenko, E., J. Duncan, and N. Kanwisher, 'Broad domain generality in focal regions of frontal and parietal cortex'. *Proceedings National Academy Sciences USA*, 2013. 110: 16616–21.

Differences between chimpanzee and human brain:

Avants, B. B., P. T. Schoenemann, and J. C. Gee, 'Lagrangian frame diffeomorphic image registration: morphometric comparison of human and chimpanzee cortex'. *Medical Image Analysis, 2006*. 10: 397–412.

Remapping factors and processing power:

Passingham, R. E. and J. B. Smaers, 'Is the prefrontal cortex especially enlarged in the human brain allometric relations and remapping factors'. *Brain Behavior Evolution*, 2014. 84: 156–66.

Chapter 6: Deciding

Prefrontal cortex and considering alternative choices:

Boorman, E. D., T. E. Behrens, and M. F. Rushworth, 'Counterfactual choice and learning in a neural network centered on human lateral frontopolar cortex'. *Public Library of Science Biology*, 2011. 9: e1001093.

Abstract scale of rewards:

Grabenhorst, F., et al., 'A common neural scale for the subjective pleasantness of different primary rewards'. *NeuroImage*, 2010. 51: 1265–74.

Course of prefrontal activation during learning of a sequence:

Toni, I., et al., 'The time-course of changes during motor sequence learning: a whole-brain fMRI study'. *NeuroImage*, 1998. 8: 50–61.

Activation when a sequence has become automatic:

Lungu, O., et al., 'Striatal and hippocampal involvement in motor sequence chunking depends on the learning strategy'. *Public Library of Science One*, 2014. 9: e103885.

Re-engagement of prefrontal cortex with attention to action:

Rowe, J., et al., 'Attention to action in Parkinson's disease: impaired effective connectivity among frontal cortical regions'. *Brain*, 2002. 125: 276–89.

Scanning internally-generated decisions:

Rowe, J. B. et al., 'Action selection: a race model for selected and non-selected actions distinguishes the contribution of premotor and prefrontal area'. *NeuroImage*, 2010. 51: 888–96.

Prospective activation in the dorsal prefrontal cortex:

Pochon, J. B. et al., 'The role of the dorsolateral prefrontal cortex in the preparation of forthcoming actions: an fMRI study'. *Cerebral Cortex*, 2001. 11: 260–6.

Scanning the Tower of London:

Rowe, J. B., et al., 'Imaging the mental components of a planning task'. *Neuropsychologia*, 2001. 31: 315–27.

Effect of prefrontal lesion on planning in an architect:

Goel, V. and J. Grafman, 'Role of the right prefrontal cortex in ill-structured planning'. *Cognitive Neuropsychology*, 2000. 17: 415–36.

Imagining a future reward:

Hakimi, S. and T. A. Hare, 'Enhanced neural responses to imagined primary rewards predict reduced monetary temporal discounting'. *Journal of Neuroscience*, 2015. 35: 13103–9.

Effect of ventromedial lesions on performance of the Iowa Gambling Task:

Bechara, A., H. Damasio, and A. R. Damasio, 'Emotion, decision making and the orbitofrontal cortex'. *Cerebral Cortex*, 2000. 10: 295–307.

Activation in the ventromedial prefrontal cortex during risky choices:

Fukui, H., et al., 'Functional activity related to risk anticipation during performance of the Iowa Gambling Task'. *NeuroImage*, 2005. 24: 253–9.

Reduced anticipatory SCR before gamblers make risky choices:

Goudriaan, A. E., et al., 'Psychophysiological determinants and concomitants of deficient decision making in pathological gamblers'. *Drug Alcohol Dependency*, 2006. 84: 231–9.

Activation in ventromedial prefrontal cortex in adolescent boys when they consider the feelings of others:

Sebastian, C. L., et al., 'Neural processing associated with cognitive and affective theory of mind in adolescents and adults'. *Social Cognitive and Affective Neuroscience*, 2012. 7: 53–63.

Effect of early ventromedial lesions on behaviour:

Anderson, S. W., et al., 'Impairments of emotion and real-world complex behavior following childhood- or adult-onset damage to ventromedial prefrontal cortex'. *Journal of the International Neuropsychological Society*, 2006. 12: 224–35.

Utilitarian reasoning in patients with ventromedial prefrontal lesions:

Thomas, B. C., K. E. Croft, and D. Tranel, 'Harming kin to save strangers: further evidence for abnormally utilitarian moral judgments after ventromedial prefrontal damage'. *Journal of Cognitive Neuroscience*, 2011. 23: 2186–96.

Lack of empathy after damage to prefrontal cortex:

Shamay-Tsoory, S. G., et al., 'Characterization of empathy deficits following prefrontal brain damage: the role of the right ventromedial prefrontal cortex'. *Journal of Cognitive Neuroscience*, 2003. 15: 324–37.

Chapter 7: Checking

Readiness potential starts before awareness on the Libet task:

Libet, B., et al., 'Time of conscious intention to act in relation to onset of cerebral activity (readiness-potential). The unconscious initiation of a freely voluntary act'. *Brain*, 1983. 106: 623–42.

Enhancement in the Pre-SMA on the Libet task:

Lau, H. C., et al., 'Attention to intention'. *Science*, 2004. 303: 1208–10.

Multi-voxel analysis predicts which movement people are going to make:

Bode, S., et al., 'Tracking the unconscious generation of free decisions using ultra-high field fMRI'. *Public Library of Science One*, 2011. 6: e21612.

Degree of enhancement and timing of awareness:

Lau, H. C., R. D. Rogers, and R. E. Passingham, 'On measuring the perceived onsets of spontaneous actions'. *Journal of Neuroscience*, 2006. 26: 7265–71.

Activation in the anterior cingulate cortex during monitoring:

Carter, C. S., et al., 'Parsing executive processes: strategic vs. evaluative functions of the anterior cingulate cortex'. *Proceedings National Academy Sciences USA*, 2000. 97: 1944–8.

Activation in the anterior cingulate cortex and awareness of errors:

Orr, C. and R. Hester, 'Error-related anterior cingulate cortex activity and the prediction of conscious error awareness'. *Frontiers in Human Neuroscience*, 2012. 6: e00177.

Re-engagement of prefrontal cortex after an error:

Kerns, J. G., et al., 'Anterior cingulate conflict monitoring and adjustments in control'. *Science*, 2004. 303: 1023–6.

Effect of anterior cingulate lesions on adaptation after an error:

Maier, M. E., et al., 'Impaired rapid error monitoring but intact error signaling following rostral anterior cingulate cortex lesions in humans'. *Frontiers in Human Neuroscience*, 2015. 9: e00339.

Videos of deceit and activation in rostral cingulate cortex:

Grèzes, J., C. Frith, and R. E. Passingham, 'Brain mechanisms for inferring deceit in the actions of others'. *Journal of Neuroscience*, 2004. 24: 5500–5.

Cartoons and activation in rostral cingulate cortex:

Amodio, D. M. and C. D. Frith, 'Meeting of minds: the medial frontal cortex and social cognition'. *Nature Reviews Neuroscience*, 2006. 7: 268–77.

Mirror system versus mentalizing:

De Lange, F. P., et al., 'Complementary systems for understanding action intentions'. *Current Biology*, 2008. 18: 454–7.

Trait words and activation in rostral cingulate cortex:

Ochsner, K. N., et al., 'The neural correlates of direct and reflected self-knowledge'. *NeuroImage*, 2005. 28: 797–814.

Areas of the human brain that are disproportionately expanded:

Glasser, M. F., et al., 'Trends and properties of human cerebral cortex: correlations with cortical myelin content'. *NeuroImage*, 2014. 93: 165–75.

Chapter 8: Acting

Right handedness in human evolution:

Toth, N., 'Archeological evidence for preferential right handedness in the lower and middle Pleistocene and its possible implications'. *Journal of Human Evolution*, 1985. 14: 607–14.

Aphasia and apraxia:

Weiss, P. H., et al., 'Where language meets meaningful action: a combined behavior and lesion analysis of aphasia and apraxia'. *Brain Structure and Function*, 2016. 221: 563–76.

Effect of left hemisphere strokes on sign language:

Hickok, G., U. Bellugi, and E. S. Klima, 'The neurobiology of sign language and its implications for the neural basis of language'. *Nature*, 1996. 381: 699–702.

Activations during sign language and spoken language:

Emmorey, K., et al., 'How sensory-motor systems impact the neural organization for language: direct contrasts between spoken and signed language'. *Frontiers in Psychology*, 2014. 5: e00484.

Activations during imitation of gestures:

Hermsdorfer, J., et al., 'Cortical correlates of gesture processing: clues to the cerebral mechanisms underlying apraxia during the imitation of meaningless gestures'. *NeuroImage*, 2001. 14: 149–61.

Asymmetry in connections from inferior parietal cortex to Broca's area:

Caspers, S., et al., 'Probabilistic fibre tract analysis of cytoarchitectonically defined human inferior parietal lobule areas reveals similarities to macaques'. *NeuroImage*, 2011. 58: 362–80.

Asymmetry in connections from superior and middle temporal cortex to Broca's area:

Rilling, J. K., et al., 'The evolution of the arcuate fasciculus revealed with comparative DTI'. *Nature Neuroscience*, 2008. 11: 426–8.

Date and mode of evolution of spoken language:

Bolhuis, J. J., et al., 'How could language have evolved?' *PLOS Biology*, 2014. 12: e1001934.

Broca's area and artificial grammar:

Bahlmann, J., R. I. Schubotz, and A. D. Friederici, 'Hierarchical artificial grammar processing engages Broca's area'. *NeuroImage*, 2008. 42: 525–34.

Broca's area, artificial grammar and neurobiological model:

Petersson, K. M. and P. Hagoort, 'The neurobiology of syntax: beyond string sets'. *Philosophical Transactions of the Royal Society B: Biological Sciences*, 2012. 367: 1971–83.

Changes in activation with learning of a motor skill:

Floyer-Lea, A. and P. M. Matthews, 'Changing brain networks for visuomotor control with increased movement automaticity'. *Journal of Neurophysiology*, 2004. 92: 2405–12.

Automation of rhythms:

Ramnani, N. and R. E. Passingham, 'Changes in the human brain during rhythm learning'. *Journal of Cognitive Neuroscience*, 2001. 13: 952–66.

Role of cerebellum in the automation of speech:

Raichle, M. E., et al., 'Practice-related changes in human brain functional anatomy during non-motor learning'. *Cerebral Cortex*, 1994. 4: 8–26.

Cerebellum and signal for prediction error over sensory consequences:

Flament, D., et al., 'Functional magnetic resonance imaging of cerebellar activation during the learning of a visuomotor dissociation task'. *Human Brain Mapping*, 1997. 4: 210–26.

Cerebellum and tickling:

Blakemore, S. J., C. D. Frith, and D. M. Wolpert, 'The cerebellum is involved in predicting the sensory consequences of action'. *Neuroreport*, 2001. 12: 1879–84.

Basal ganglia and prediction error for reward:

Valentin, V. V. and J. P. O'Doherty, 'Overlapping prediction errors in dorsal striatum during instrumental learning with juice and money reward in the human brain'. *Journal of Neurophysiology*, 2009. 102: 3384–91.

Chapter 9: The future

MEG and imitation:

Nishitani, N. and R. Hari, 'Viewing lip forms: cortical dynamics'. *Neuron*, 2002. 36: 1211–20.

Recording from parietal cortex with a multi-electrode array:

Aflalo, T., et al., 'Neurophysiology: decoding motor imagery from the posterior parietal cortex of a tetraplegic human'. *Science*, 2015. 348: 906–10.

Simulated reaching in tetraplegia:

Chadwick, E. K., et al., 'Continuous neuronal ensemble control of simulated arm reaching by a human with tetraplegia'. *Journal of Neural Engineering*, 2011. 8: e034003.

Recording from single neurons in the hippocampus:

Quiroga, R. Q., et al., 'Invariant visual representation by single neurons in the human brain'. *Nature*, 2005. 435: 1102–7.

Computer model of microcircuit in rat somatosensory cortex:

Markram, H., et al., 'Reconstruction and simulation of neocortical microcircuitry'. *Cell*, 2015. 163: 456–92.

AlphaGo programme:

Silver, D., et al., 'Mastering the game of Go with deep neural networks and tree search'. *Nature*, 2016. 529: 484–9.

Further reading

Very short introductions

Butler, G. and F. McManus, *Psychology*. 2000, Oxford University Press: Oxford.

Foster, J. K., *Memory*. 2008, Oxford University Press: Oxford.

O'Shea, M., *The Brain*. 2005, Oxford University Press: Oxford.

Textbooks

Gazzaniga, M. S., R. B. Ivry, and G. R. Mangun, *Cognitive Neuroscience: The Biology of the Mind*. 4th Edition, 2014, Norton: New York.

Uttal, W. R., *Mind and Brain: A Critical Appraisal of Cognitive Neuroscience*. 2011, MIT Press: Cambridge.

Ward, J., *The Student's Guide to Cognitive Neuroscience*. 3rd Edition, 2013, Taylor and Francis: London.

Brain imaging

Passingham, R. E. and J. B. Rowe, *A Short Guide to Brain Imaging: The Neuroscience of Human Cognition*. 2015, Oxford University Press: Oxford.

Raichle, M. E., 'A brief history of brain mapping'. *Trends in Neuroscience*, 2009. 32: 118–26.

Reviews of specific topics in cognitive neuroscience

Amodio, D. M. and C. Frith, 'Meeting of minds: the medial frontal cortex and social cognition'. *Nature Reviews Neuroscience*, 2006. 7: 268–77.

Bechara, A., et al., 'Emotion, decision making and the orbitofrontal cortex'. *Cerebral Cortex*, 2000. 10: 295–307.

Corbetta, M. and G. L. Shulman, 'Control of goal-directed and stimulus-driven attention in the brain'. *Nature Reviews Neuroscience*, 2002. 3: 201–15.

Duncan, J., 'An adaptive coding model of neural function in prefrontal cortex'. *Nature Reviews Neuroscience*, 2001. 2: 820–9.

Friederici, A. D. and W. Singer, 'Grounding language processing in basic neurophysiological principles'. *Journal of Cognitive Neuroscience*, 2015. 19: 329–38.

Haggard, P., 'Human volition: towards a neuroscience of will'. *Nature Reviews Neuroscience*, 2008. 9: 934–46.

Hubbard, E. M., et al., 'Interactions between number and space in parietal cortex'. *Nature Reviews Neuroscience*, 2005. 6: 435–48.

Luo, L., et al., 'Ten years of Nature Reviews Neuroscience: insights from the highly cited'. *Nature Reviews Neuroscience*, 2011. 10: 18–28.

Malach, R., et al., 'The topography of high-order human object areas'. *Trends in Cognitive Sciences*, 2002. 6: 176–84.

Passingham, R. E., et al., 'The anatomical basis of functional localization'. *Nature Reviews Neuroscience*, 2002. 3: 606–16.

Rushworth, M. E., et al., 'Action sets and decisions in the medial frontal cortex'. *Trends in Cognitive Sciences*, 2004. 8: 410–17.

Sitt, J. D., et al., 'Ripples of consciousness'. *Trends in Cognitive Sciences*, 2013. 17: 552–4.

Spiers, H. and E. A. Maguire, 'Decoding human brain activity during real-world experiences'. *Trends in Cognitive Sciences*, 2007. 11: 356–65.

Index

C

D

E

F

G

N

O

P

R

S

T

V

W

Z